Poultry Breeding and Management

by Professor James Dryden

with an introduction by Jackson Chambers

This work contains material that was originally published in 1916.

This publication is within the Public Domain.

*This edition is reprinted for educational purposes
and in accordance with all applicable Federal Laws.*

Introduction Copyright 2018 by Jackson Chambers

The World's Largest Selection of Vintage Poultry Books

www.VintagePoultry.com

Introduction

I am pleased to present yet another title on Poultry.

The work is in the Public Domain and is re-printed here in accordance with Federal Laws.

As with all reprinted books of this age that are intended to perfectly reproduce the original edition, considerable pains and effort had to be undertaken to correct fading and sometimes outright damage to existing proofs of this title. At times, this task is quite monumental, requiring an almost total "rebuilding" of some pages from digital proofs of multiple copies. Despite this, imperfections still sometimes exist in the final proof and may detract from the visual appearance of the text.

I hope you enjoy reading this book as much as I enjoyed making it available to readers again.

Jackson Chambers

LADY MACDUFF—The First 300-Egg Hen:

This Oregon Agricultural College hen has demonstrated the high egg-producing possibilities of the domestic hen by laying 303 eggs in 12 months, 512 eggs in 24 months, and 679 eggs in 36 months.

PREFACE

THERE need be no apologies for new poultry books. The industry is important, the poultry constituency large, and one poultry book representing the finding of one author would hardly be presumed to meet all demands. In these days of progress in the science, if it may be so called, of poultry husbandry, it is imperative that new compilations be made and new books published at frequent intervals, that the poultry keeper may receive the benefit of early knowledge of new discoveries.

The remarkable development of poultry culture during the past two decades is one of the outstanding features of American agriculture. Twenty years ago the possibilities of poultry-keeping as an industry were scarcely dreamed of. While it does not yet receive the consideration it deserves—far from it—nevertheless it has made immense gains both in popular recognition and in production. This has been brought about by a better realization of the productive value of the hen. The idea of "fuss and feathers," long associated with the keeping of fowls, has gradually given way to the idea of a poultry industry whose first and only business, as an industry, is the production of eggs and meat.

With the development of the industry, there has been a growing demand for information dealing with practical problems of production. The poultry producer has his full share of problems. It must be confessed that the available literature has been insufficient and fragmentary. This lack, however, is being rapidly filled, and, as a result, in all parts of the country, there are now examples of suc-

PREFACE

cessful poultry farms; not that the special poultry farm is by any means a true measure of the poultry industry, for the industry is, and probably always will be, largely a business for the general farmer, but that the success of special poultry-keeping is a measure of the advance that has been made in the solution of practical poultry problems.

This book, therefore, has been prepared that it may add to the available poultry literature; not that it may supplant other books, nor that it should be the last word on the subject. The author is fully conscious of its imperfections; but, to every student of poultry culture, and to every poultry farmer, he earnestly hopes that it may bring some helpful message.

JAMES DRYDEN.

Corvallis, Oregon.

CONTENTS

Chapter		Page
I.	Historical Aspect	1
II.	Evolution of Modern Fowl	11
III.	Modern Development of Industry	19
IV.	Classification of Breeds	24
V.	Origin and Description of Breeds	30
VI.	Principles of Poultry Breeding	61
VII.	Problem of Higher Fecundity	92
VIII.	Systems of Poultry Farming	138
IX.	Housing of Poultry	160
X.	Kind of House to Build	187
XI.	Fundamentals of Feeding	210
XII.	Common Poultry Foods	237
XIII.	Methods of Feeding	249
XIV.	Methods of Hatching Chickens	281
XV.	Artificial Brooding	320
XVI.	Marketing Eggs and Poultry	333
XVII.	Diseases and Parasites of Fowls	375

ILLUSTRATIONS

	Page
Lady Macduff. The first 300-egg hen	*Frontispiece*
Jungle fowl cock (*Gallus bankivus*)	3
Jungle fowl hen (*Gallus bankivus*)	4
White Leghorn cockerel	31
Black Minorca male	32
Barred Plymouth Rock cockerel	39
Barred Rock hen, showing fine barring	40
White Wyandotte hen	42
White Wyandotte cock	43
Rhode Island Reds	44
White Orpington hen	48
Light Sussex	49
Speckled Sussex	50
Domesticated	51
Faverolle hen	51
Three breeds of different types winning in Australian laying competitions	53
Good utility Barred Plymouth Rock cockerel	55
Light Brahmas	56
Buff Cochin hen	57
Le Mans—A special French meat breed	58
Points of the fowl	59
La Fleche	60
Variation the opportunity of the breeder	64
Result of crossing White Wyandotte and Black Minorca, showing barring	67
Breed improvers. Pedigreed cockerels, from stock with records of over 200 eggs in a year	69
Barred Plymouth Rock male. Son of a 218-egg hen	71
A good type of breeder from 200-egg stock	72
Barred Plymouth Rock male, Oregon Station	73
Result of breeding for a fancy point	74
Barred Rock and White Leghorn first cross, male	78

ILLUSTRATIONS

	Page
Barred Rock and White Leghorn, first cross, female	78
Barred Rock and White Leghorn first cross. Flock showing dominant white	79
White Wyandotte-Black Minorca male, first cross, with white plumage and rose comb	80
The recessive color barring	81
Oregon Station hen C543. An exceptional layer though inbred	89
Like begets like	92
Like does not always beget like	93
Oregon Station hen D18, 271 eggs in a year	93
Barred Rock hen A78, record 212 eggs	94
A good Plymouth Rock head with the stamp of vigor	95
Barred Plymouth Rock hen, 65, 218 eggs	96
Daughter of 65, laid 218 eggs. Granddaughter, laid 221 eggs	97
A mother of high producers, A122 laid 259 eggs in a year	98
A family of high producers, daughters of A122	99
A77, 214 eggs. A producer of good layers	100
Daughters of A77	101
Granddaughters of A77	102
A79, 219 eggs. A good layer and breeder of good layers	104
Granddaughters of A79	105
White Leghorn hen O34, 229 eggs in first year	107
Daughters of O34	107
Oregona—White Leghorn hen. Record of more than 1,100 eggs	108
Hen B42 laid 834 eggs in four years	113
Hen A60 laid 816 eggs in four years	113
Belle of Jersey, 649 eggs in three years	116
Queen Utana, 816 eggs in five years	116
Rose-Comb Brown Leghorn hen, 442 eggs in two years	116
Three Cornell long-distance layers	117
Lady Macduff, taken day after she laid her 303d egg	117
The 303d egg of Lady Macduff	117
Lady Macduff and 10 daughters	118
Pedigree of Lady Macduff	118
Lady Macduff in full plumage in her second year	119
Son of Lady Macduff	119
Daughter of Lady Macduff	119
Mother of Lady Macduff	120
Lady Showyou	121
C543 at end of first 12 months' laying	122

ILLUSTRATIONS

	Page
Oregon hen C543, 291 eggs	122
Highest record hen at the Missouri 1913-14 competition	122
The head indicating laying quality	123
Breeder of poor layers, 20 eggs in a year	124
White Leghorn hen, laid 1 egg in a year	124
A good layer from poor laying stock	125
Poor layers from good laying stock	125
Barred Plymouth Rock hen laid 74 eggs in a year	126
Another view of Oregona	126
White Leghorn hen, 242 eggs in first year	126
Hen E248, 302 eggs. Daughter of C516	127
White Leghorn hen C516, 267 eggs in a year	127
Two poor layers	128
Utah Station Wyandottes	128
New Zealand White Leghorns	129
White Wyandottes averaged 208.5 eggs in Storrs contest	129
White Leghorns averaged 208.8 eggs in Storrs Competition	130
Winning pen in the Panama Pacific International Egg-laying Competition	130
The long and the short way, in breeding for eggs	131
Record of a flock of 43 fowls at the Oregon Station for two years	131
Good fall and winter producers the best layers	132
The first layers the best layers	133
Inheritance of egg production	134, 135
Egg organs of the hen	136
Poultry keeping and dairying	139
An Oregon fruit and poultry farm	141
A California poultry and fruit farm	141
1,000 pullets in prune orchard	142
Eggs and peaches from the same ground	143
Free range colony system at Petaluma, California	145
Petaluma farm of 120 acres and 6,000 hens	146
Cleaning out the houses on a Petaluma farm	147
Land unfit for cultivation is used	148
2,000 hens on 3 acres	148
Exclusive poultry farming on the intensive system	149
4,000 hens on 4 acres	150
The intensive plan	151
Backyard egg farming	152

ILLUSTRATIONS

	Page
A poultry yard may be made an attractive feature of the backyard	153
A backyard house in which 25 hens averaged 188 eggs	154
Another backyard system	155
A plan for backyard poultry keeping	156
The first and not the worst poultry house	163
About the worst poultry house that was ever built	164
An unsatisfactory poultry house	165
A boy with a "safe" horse and "spring" wagon gathers the eggs	176
A rear view of the Missouri house, showing ventilation	171
The Missouri Poultry Station house	172
The Oregon Station's first open front colony house	177
The improved Oregon Station portable house	178
The Oregon Station pullet-testing yards	179
Special breeding yards, Oregon Station	180
A colony house at the Utah Station	181
A scratching shed is an advantage where the house room is limited	182
A cheap shed for fowls	183
Colony houses used on a Rhode Island poultry farm	184
Piano boxes utilized for hen houses	185
Stationary 100-hen house. Oregon Station	188
Curtain-front house	189
The nests arranged under the dropping platform	197
A good broody coop	198
The Oregon Station trapnest	200
Taking Lady Macduff from the trapnest when she laid her 303d egg	201
Portable fence	204
Showing how fence may be constructed	205
A balanced ration	219
The relative amounts of ash, fat, protein and water in eggs	220
Balanced ration for one hen for a year	226
Digestive organs of the fowl	235
Chickens threshing their own grain	238
Oregon Station outdoor dry food hopper	255
A good roaster	275
Feeding battery for fattening	277
Feeding Station	279

ILLUSTRATIONS

	Page
Parts of a fresh egg	282
The beginning of the end of incubation	284
The young graduate	284
Nests used for sitting hens	291
The Oregon Station combination hatching and brooding coop	293
A new brood coop	294
Plan of hen brood coop	295
Hen brooding at Oregon Station	296
Brood coop made out of a shoe box	297
Brooding coops on a Rhode Island farm	298
Division of poultry labor at Petaluma	300
Chicks loaded onto the wagon	301
After traveling two miles the chicks were put into this brooding house	302
Oregon Station incubator house	303
Interior of Oregon Station incubator house	304
A 150-egg incubator	305
A 250-egg incubator	306
Hot water jug brooder	325
Continuous brooding system	326
Room or stove brooding; a night scene	327
A stove brooder with hover	328
A stove brooder showing hover and different parts	329
Room brooding, with oil or gas heater outside of room	330
Flock of 8,000 young pullets	330
Cornell gasoline brooder	331
Terra cotta brooder	332
A 12-dozen crate which may be used for shipping eggs	338
A roaster in a parcel post package	339
A parcel post package showing eggs wrapped	339
The rural mail carrier takes the eggs from the farm	340
Commercial egg candling	352
A kerosene lamp set inside of a box makes a good tester	353
Instead of a kerosene lamp, an electric light bulb may be used	354
A fresh egg—note small air space	355
A stale egg—note large air space	355
Cans of frozen eggs	361
Poultry demonstration car	367
Unloading a Nebraska carload of poultry at San Francisco	368
Dry picking, dry cooling, and dry-packed poultry	370

ILLUSTRATIONS

	Page
Dressed capon	373
An expert caponizer	373
A bad case of roup	381
Normal hen's ovary	387
Diseased ovary	387
Two white diarrhœa chicks	389
Taking blood sample for white diarrhœa test	391

POULTRY BREEDING AND MANAGEMENT

CHAPTER I

HISTORICAL ASPECT

Present races of fowls were domesticated, or reclaimed from the wild state, away back about the time that man was learning the rudiments of civilization. When man himself became "tame," he set about taming the wild things of the forest and the plain, in order that they might better supply his needs for food, for raiment, and for labor. "A bird in the hand is worth two in the bush," was undoubtedly the impelling motive that led to the domestication of the wild fowl. Savages were content to depend upon the hunt for their daily food supply. Centuries after the ancient peoples of Asia had domesticated the fowl, the Indians on this continent had failed to domesticate the turkey, which is now the most highly prized bird for food, and possibly the most highly valued of any kind of animal food.

Civilized man desired a more certain food supply, however, than that of the hunt. To exercise his God-given dominion over the earth, man had to bring to his assistance plants and animals that hitherto existed only in the wild state. With domestication, came improvement in productive qualities. The eggs of the wild fowl had no other use than reproduction. She laid a few eggs and hatched them. There was no demand for them for human food, or for use in arts and manufactures. The wild ancestor of our domestic hen laid probably a dozen or twenty eggs a year. The difference between that and eight or ten dozen represents the achievement of centuries of poultry culture.

The purpose of domestication was undoubtedly utility. There is no evidence to show that fowls were domesticated for any fancied or peculiar appearance. There were other birds that appealed more to the æsthetic. There were various species of the pheasant family, of gorgeous plumage and proud carriage—all have remained practically in their natural state. If the ancients were looking for something to please the eye or the fancy, some of these would have suited their purpose better than the fowl. Our present breeds of fowls, however much some of them may be embellished with colors and shapes that appeal to our fancy and command our admiration, are without "pride of ancestry," so far as the original jungle fowl conformed to our present-day standard of beauty. But it is not surprising that after thousands of years of poultry keeping we have now some breeds that have been developed along fancy lines entirely.

There might have been another object besides utility in domesticating the fowl. Semi-barbarous peoples of the Orient were, and still are, much addicted to the sport of cock-fighting, and the fighting qualities of the jungle fowl may have appealed to them more than any possible use they could make of the fowl as a source of food supply.

The fact that there has been great improvement in meat and egg production, however, is pretty strong evidence that usefulness was the impelling motive in the domestication of fowls and in their breeding through all the centuries since they were weaned away from their natural state.

Origin.—It is generally agreed among naturalists that our present races of domestic fowls are descended from a wild jungle fowl of India. The Orient has given to the world the fowl as well as many of our domestic animals. There are four species of jungle fowls from which it is claimed by different authorities that domestic fowls were descended, namely, *Gallus bankiva, G. Sonneratii, G. Stanleyii, G. Varius* (or *furcatus*). While there has been much discussion and difference of opinion, it is generally conceded that the evidence points to the *Gallus bankivus* as the original progenitor. This species is a Bantam-sized fowl, patterned much after the Red Game of our day. The male Bankivus has the color and carriage of the Game.

"Specimens of this fowl," says Mr. Dixon, "were brought from the island of Java and deposited in the museum of Paris. They inhabit the forests and borders of woods, and are exceedingly wild. On examining the species, it will be found to exhibit many points of resemblance with our common barnyard fowls of the smaller or middling size. The form and color are the same, the comb and wattles are smaller, and the hen so much resembles the common hen that it is difficult to distinguish it except by the less erect slant of the tail. The rise of the tail is much more apparent in the male, but it may be observed that in all wild species known, the tail does not rise so high above the level of the rump, nor is it so abundantly provided with covering feathers as in the common birds. Feathers which fall from the neck over the top of the back are, as in other fowls, long and with divided plumelets or braids, the feathers widening a little and being rounded. The colors of the plumage are exceedingly brilliant. The head, the neck, and all the long

JUNGLE FOWL COCK (*Gallus bankivus*)
Reproduced from Carnegie Institution. Publication No. 121, 1909, by Charles S. Davenport.

feathers of the back which hang over the rump are of a shining, flame-colored orange; the top of the back, the small and middle coverts of the wings are of a fine maroon purple; the coverts of the wings are black, tinged with irridescent green; the quill feathers of the wings are russet red on the outer and black on the inner edges; the breast, belly, thighs, and tail are black and tinged with irridescent green; the comb, cheek, throat, and wattles are of a more or less vivid red; the legs and feet are grey and furnished with strong spurs, the iris of the eye yellow.

"The Bankiva hen is smaller than the cock; and her tail is also a little horizontal and vaulted; she has a small comb, and the wattles are very short; the space around the neck, as well as the throat, is naked; on this space are some small feathers, distinct from each other, through which the red skin can be seen; the breast and belly are light bay or fawn yellow, and on each feather is a small, clear ray along the side of the middle rib or stem; the feathers of the base of the neck are long, with disunited braids, or plumelets, of a black color in the middle and fringed with ochre yellow; the back, the coverts, the wings, the rump and the tail are earthy grey marked with numerous black zigzags; the large feathers of the wing are ashy grey."

The Bankivus inhabits northern India and is found in

JUNGLE FOWL HEN (*Gallus bankivus*)
Reproduced from Carnegie Institution. Publication No. 121, 1909, by Charles B. Davenport.

the Himalayan mountains at an altitude of 4,000 feet; higher up other species of wild fowl are found. It also inhabits Burma, the Malay Peninsula and the Philippine Islands, and the Island of Java.

The evidence in support of a common origin of all races of fowls comes largely from Darwin. While Darwin was inclined to a belief in a common origin and saw nothing impossible in this theory, at the same time there are indications in his writings that he thought it barely possible that some varieties of fowls might have been descended from a different species, now possibly extinct. On the other hand, some poultry fanciers took issue with Darwin and proclaimed it impossible that all domestic fowls could have been descended from one parent source.

Darwin based his conclusions largely on his own experiments, and while, as he himself confesses, the evidence may not be conclusive, it is the best evidence that we have, and we give here the substance of it. The evidence pointed to the Bankivus as the progenitor of all fowls, first, because it mated with the tame fowl and produced offspring, while the other species mentioned never or rarely crossed. Darwin dwells with considerable detail on this fact as an argument in favor of the Bankivus. Sometimes, however, different species of animals will mate together and produce offspring, but the progeny called hybrids are barren or unfertile. The mule is usually cited in illustration of this fact. He is the product of two distinct species of animals, the proof of which is the fact that he is barren. The horse and the ass therefore could not have had a common origin.

Darwin, and later others, not only found that the *Gallus bankivus* freely mates with our domestic fowl, but that the offspring are fertile and breed successfully. These experiments strongly impressed Darwin with the belief that *Gallus bankivus* is the original progenitor of domestic fowls.

Experiments along another line pointed in the same direction. Students of heredity know that crossing and intercrossing breeds and varieties cause reversion, or a breeding back to remote ancestors. Why the likeness of some ancient ancestor through the act of crossing different breeds should suddenly reappear in the offspring after having apparently disappeared from the face of the earth centuries ago is one of the enigmas of breeding. Following up the clue of reversion, Darwin found what he claimed to be strong evidence pointing to the *Gallus bankiva* as the original ancestor of our fowls. He says that Game, Malay, Cochin, Bantam and Silkies, when crossed, revert to the Bankiva. In crossing the Black Spanish and White Silkie, he found that the offspring were all black, except one cock which resembled *Gallus bankiva* so strongly that he said: "It was a marvelous sight to compare this bird with *Gallus bankiva* and then with its father."

He declared further that the color of the golden and silver Pencilled Hamburgs pointed to their ancient progenitors. "This may be in part explained by direct reversion to the parent form, the Bankiva hen, for this bird has all its upper plumage finely mottled." Remarkable, is it not, that after two or three thousand years of breeding away from the wild fowl, it is possible in crossing to trace in the color of plumage and shape and carriage of the offspring the descent of the wild fowl to our present modern breeds. And yet to scientists such as Darwin, mute testimony of this nature may be more conclusive than the written word.

Darwin's findings in regard to the common origin of the domestic fowl may be summarized as follows:

1. The domestic fowls mate freely with *G. bankiva*.
2. They mate very rarely with any other species.
3. The Bankiva hybrids are fertile.
4. The hybrids of other species are not fertile.

HISTORICAL ASPECT

He argued in favor of but one origin, namely *G. bankiva*.

He explains in the following words how the changes in the fowl have come about, and how it is reasonable to believe that all the breeds have descended from one parent source:

" . . . from the occasional appearance of abnormal characters, though at first only slight in degree; from the effects of the use and disuse of parts; possibly from the direct effects of changed climate and food; from correlation of growth; from occasional reversions to old and long-lost characters; from the crossing of breeds when more than one had once been formed; but, above all, from unconscious selection carried on during many generations."

While the views and conclusions of Darwin were generally those of all naturalists, there were others, including poultry writers and fanciers, who took strong grounds against them. His conclusions were published in the year 1867. It is worthy of note that a gentleman from whom he got much of his poultry information, and whom he frequently quotes in his book, later (1885) took issue with his conclusions that all domestic fowls came from *Gallus bankivus*. This man was Mr. W. B. Tegetmeier, F.Z.S., a noted poultry author and authority in England, who was associated with Darwin in some of his experiments. Mr. Tegetmeier gives it as his opinion that the different species of wild Galli will interbreed, and then he says:

"But it is with regard to the Eastern Asiatic type of fowl (absurdly known as Cochins and Brahmas) that my doubts as to the descent from the *G. ferrugineus* (Bankivus) are strongest. We have in the Cochin a fowl so different from the ordinary domestic birds that when first introduced the most ridiculous legends were current respecting it. Putting these on one side, we have a bird with many structural peculiarities that could hardly have been induced by domestication. Thus, the long axis of the occipital foramen in the Cochin is perpendicular, in our old breeds horizontal, a difference that could never have been bred for, and which it is difficult to see could be correlative with any other change. The same may be said respecting the deep sulcus or groove up the center of the frontal bone.

"The extraordinary diminution in the size of the flight feathers and that of the pectoral muscles could hardly have been the result of human selection and careful breeding, as the value of the birds as articles of food is considerably lessened by the absence of flesh on the breast. Nor is the extreme abundance of fluffy, soft body feathers a character likely to be desired in a fowl. The vastly increased size may have been a matter of selection, although, as the inhabitants of Shanghai feed their poultry but scantily, and, according to Mr. Fortune, mainly on paddy of unhusked rice, it is not easy to see how the size of the breed was obtained if, as is generally surmised, it arose from the little jungle fowl.

"Taking all these facts into consideration, I am induced to believe that the birds of the Cochin type did not descend from the same species as our game fowl."

Mr. Edward Brown expresses his opinion as follows:

"To sum up, therefore, it may be taken that with the domestic fowl, as with many other natural forms of life, we can go so far back, but no further. The probability is that, as in the case of dogs, all the varieties of fowls do not owe their origin to any one species, at any rate of those now extant, and that we must look to another progenitor than the *G. ferrugineus* for several of the later introduced races, more especially those from China."

Such, briefly stated, is the argument, pro and con, as to the common origin of the domestic fowl. It may be enough for us to know that we have the chicken that lays the eggs and feeds the world. In the jungles of farther India a wild fowl is scratching and cackling to-day as its ancestors did three thousand years ago. It breeds pure without any standard of excellence, and lays the same number of eggs as its ancestor did before the Christian era. It crows at the midnight hour, but it shuns the society of man. It is pure-bred because it has the same characteristics as a thousand ancestors have had. While it revels in the jungle and abhors the sight of man it has millions of relatives living useful lives, ministering to the wants of man, and on two continents producing yearly a billion dollars worth of poultry food-products, just because away back three thousand years ago a few of its ancestors were caught and robbed of the freedom of the jungle. What a triumph domestication of the fowl has been! What a mint of money it has coined since it gave up its freedom in the wild and became a part of civilization.

Antiquity of Domestic Fowl.—Let us now consider briefly the antiquity of the fowl. It is not possible to give dates; it is not even possible to give the century when the fowl was domesticated. It is known from New Testament scripture that cocks and hens existed two thousand years ago. There is no reference to them in the Old Testament; but we find the egg spoken of by Job in these words: "Is there any taste in the white of an egg?" As to the kind of egg we are left in doubt. That fowls were under domestication two thousand years ago there is no doubt; that they existed several hundred years before that, there is authentic proof; how much longer must remain largely a matter of conjecture.

In tracing the antiquity of the hen, the following facts have been mentioned: When Peter denied the Savior the cock crowed thrice. That establishes the origin of the fowl before the Christian era. Mention is made of cock-fighting in the Codes of Mann, a thousand years or more before Christ. A Chinese encyclopedia, 1400 years B.C., mentions the fowl. In the religion of Zoroaster the cock figures as a sacred bird. Figures on Babylonian cylinders show that there must have been fowls in the seventh century B.C. Homer makes no mention of fowls, 900 B.C., but they are referred to in the writings of Theognis and Aristophanes about 500 B.C. The ancient Egyptian monuments are silent about the fowl, though flocks of tame geese are shown.

CHAPTER II

EVOLUTION OF MODERN FOWL

By what process, then, has the small jungle fowl, producing little meat and few eggs, been converted into the Brahma and the Leghorn of great meat- and egg-producing qualities? What brought about the change in the fowl that enables the poultryman of to-day to gather ten dozen eggs a year instead of one dozen or a dozen and a half, which was the order of the hen-day at the birth of chicken civilization? By what miracle has the meat on the fowl's skeleton been multiplied six times? Whence have come the various colors of feather, the top-knot, the feather legs, and tails 20 feet long?

There has been abundant opportunity in some three thousand years for the type and characteristics of the jungle fowl to be largely lost in the evolution of newer and better races of fowls. If the modern horse is descended from an animal not much larger than a Jack rabbit, why not a Brahma from a Bantam-sized fowl? We must disabuse our minds of the idea that poultry-keeping is a modern institution. It is idle to repeat that the fowl we see to-day on the farms and in the backyards are the product of the past fifty or even hundred years. It has taken hundreds or thousands of years to bring them to the stage of perfection that we now have them. Harrison Weir in "The Poultry Book" says on this point: "Nearly all our modern methods are only the old ones re-substituted, even that of the incubator. In the olden time they kept fowls and bred chickens with a greater certainty and in better health than many of the now professed poultrymen of the day.'' Columela two thousand years ago described the fowl, and less than a hundred years ago the description of the farm fowl of England corresponded in nearly every respect to those described by Columela.

Darwin says that "not only careful breeding but actual selection was practiced during ancient periods and by barely civilized races of men." It is pointed out that in early times there were different breeds of fowls. Six or seven are mentioned as being kept by the Romans at the commencement of the Christian era. As proof that the work of selection has not all been confined to civilized people, it is shown that the semi-barbarous people of the Philippine Islands about fifty years ago had no less than nine varieties of the game fowl. In the fifteenth century several breeds were known in Europe, and in China about the same period seven kinds were named. Finally, Darwin says: "Will it then be pretended that those persons who in ancient times and in semi-civilized countries took pains to keep breeds distinct, and who therefore valued them, would not occasionally have destroyed inferior birds and occasionally have preserved their best birds? That is all that is required."

The work of modern times has not been so much to maintain the original purity of races as it has been to make new breeds and varieties by fusing pure ancient races, to what purpose will be discussed in a later chapter. Present-day poultry breeders and livestock breeders breed for uniformity. Their skill is exercised in producing uniformity in the stock. A standard of excellence is set up, and the nearer their fowls or animals approach that standard the greater value they have in the market for breeding purposes. The poultry shows demand a certain standard, and this calls for uniformity. But uniformity does not permit of improvement or progress. It is clear that if two thousand years ago a standard of excellence corresponding to the type of jungle fowl had been set up, there would have been no such improvement in the breeds as we have to-day. The improvement did not come from breeding to a standard of uniformity. The improvement came about rather by variation.

There can be no improvement without variation. In other words, if like always followed like, improvement would be impossible. "Like begets like" is not literally true, a fact for which some of us humans may have regrets; but on the whole the human race has improved in many particulars since the days of our barbarous ancestors, largely because of this great law of heredity, the tendency to vary. The exceptional individual appears in the flock—an exceptional bird from average birds—and this is called variation. A large bird from small parents may breed a strain or variety of large fowls. A small bird from large parents may breed a variety of small birds. All plants and animals vary, and it is in taking advantage of this factor that our fowls in two or three thousand years have been bred up to a higher utility; or rather, it is one of the most potent factors. Fowls may be induced to vary in different ways. Changes in climatic conditions; changes in food and care, and crossing of different breeds, all have an influence toward greater variation. Pedigree or ancestry is a valuable asset, but sometimes the law of variation breaks into the preserve and takes captive this asset and gives us something more valuable. The old law of breeding was that pedigree was everything, and if a phenomenal individual should appear, he would quickly disappear, his offspring would be reduced to the general average. The new Mendelian view is that the phenomenal individual may breed pure; that he may defy his pedigree and ancestry and breed a superior race.

Selection.—Variation is effective through selection. Variation is responsible for the exceptional individual; selection is responsible for preserving it. Darwin says there are two kinds of selection, one he calls "unconscious," the other "methodical." To the former he credits, in large part, the evolution of the fowl. The fancier who in spirit of rivalry tries to excel his neighbor by breeding from his best bird, without any attempt to establish a new breed or to preserve some new characteristic, or improve the breed, is practicing unconscious selection. There has always been a standard of excellence, written or unwritten, and fanciers or poultry breeders have been unconsciously following it throughout the centuries. Methodical selection, on the other hand, has to do with the making of breeds and the fixing of new and desirable characteristics. This pre-supposes a knowledge on the part of the breeder of the principles of breeding. According to Darwin, unconscious selection has done more for the improvement of fowls because it has been at work longer than methodical selection.

The breeder who follows methodical selection is constantly on the lookout for new and valuable characteristics. He is not satisfied with following a standard of excellence; he sets up a new and higher standard; he believes in progress. While the man who is content to beat his neighbor in the show room and discards everything in his breeding pen that does not conform to the standard set up for prize-winners, the man who follows methodical selection would often achieve his highest purpose by breeding for characteristics or type that would have no standing in the show room. He is looking for "sports" or "mutants" along certain lines and when they appear he makes them the basis of his breeding operations.

Whether the improvement or evolution of the fowl is due more to one or the other method of selection, it would have been clearly impossible to evolve the fowl as we now have it if, in the early centuries, an arbitrary standard had been set up and all breeding made to follow along that line.

Causes of Variation.—A more abundant food supply undoubtedly accounts for many of the differences between the wild fowl and the modern tame fowl. The wild fowl varies little. It breeds true century after century, but under domestication it rapidly evolves new characteristics. Egg production depends upon a steady supply of good food. This would not be secured in the wild state. The effect of domestication has been at once to increase fecundity. The wild fowl laid a dozen or possibly two dozen eggs in a year; the tame fowl now lays ten times as many. A change of climate and a change of soil induce variation and increase vigor, and these have been potent factors, doubtless, in increased egg production. An abundant food supply operates in the same direction. Higher production came immediately into play when the fowl was put under conditions more congenial to egg production.

It is known that the wild pheasant under confinement produces twice the number of eggs that she produces in the wild state. Mr. Simpson of the Oregon State Game Farm gets an average of about sixty eggs a year from his China pheasants, and he has known them to lay a hundred, while in nature they lay but two sittings of about 13 eggs each. Another pheasant raiser is reported in United States Farmers Bulletin 390 as stating that seven of his hens laid 131 eggs and then stopped, but when he put them into a fresh pen they laid 174 more.

A change therefore to congenial surroundings or environment at once gives a decided increase in production. It would seem that a large part of the increased productiveness is due not so much to selection but to improved environment. If we are to accept the United States census figures of about 80 eggs a year as the average production of the hens of the United States, it is no more of an increase over the jungle fowl's production than might reasonably be expected from better environmental conditions.

Effect of Food and Climate on Size and Meat Qualities.—An abundant supply of food, changed soil, climate, and other conditions relating to environment, would not only increase production of eggs, but would tend to produce variations in the size and meat qualities. Abnormal characteristics, sports or mutations, would frequently result. Ancient peoples (poultry-keepers in other centuries) had no interest in standards of uniformity, and they would preserve the peculiar or abnormal birds, and new types would be evolved. A sport might be produced weighing two or three times more than the common fowl, and it is easily understood that a fowl of that size would be carefully preserved. The variations would extend to egg-laying. The poultry-keeper would find a fowl that was evidently a good layer; she would be retained and her eggs would be hatched. This kind of selection, unconscious selection, works by centuries slowly but surely.

The ancients were not influenced by prizes and high prices, but more by novelties, and it is easy to believe that the fowl of unusual or abnormal appearance was carefully preserved. So, too, it is reasonable to believe that a few out of many must have paid attention to productive qualities, and in some way—if not by trapnests, by some other method—picked out the best layers and bred from them. There is no record, so far as known, that fowls were kept to please the fancy and win prizes for fancy points, and if they were kept mainly or wholly for their economic qualities, selection must have been based on the idea of improving productive qualities. Whether we owe most to the Mutant—"the occasional appearance of abnormal characters"—or to the slow process of unconscious selection—the survival of the fittest—is a matter of speculation rather than fact.

Use and Disuse of Parts.—The use and disuse of parts has been a factor in the evolution of the fowl. How so?

The blacksmith's arm is a striking object lesson of the effect of vigorous use of a part of the body. The poultryman knows that exercise hardens the muscles of the chicken, and when a tender article of chicken meat is desired, the fowls are fattened in crates or small pens in order to keep them from exercising. Whether the qualities of tenderness in the meat could become a fixed and transmissible characteristic may be open to debate; but acquired characteristics sometimes become hereditary. The horse was originally a pacer; trotting is an acquired characteristic. Fowls were originally all sitters, but certain breeds through disuse of sitting or hatching have acquired the characteristic of non-sitting, and they breed true to that characteristic.

It is known that the wing of the tame duck has diminished in weight in proportion to size of body and legs since domestication, the tame duck being a descendant of the common wild duck. The tame duck is much larger in limb and body than its wild ancestor, and it has little or no use for its wings. By use, the leg bones have increased in size, but by disuse the wing bones have rather decreased. Under domestication, the disproportion between strength of wing and ability to fly, has become so great that a duck of the Pekin type would make a spectacle of itself if on the wide-open prairie by the use of its wings it sought to elude the pursuit of the coyote.

The same thing is true to a greater or less extent in breeds of chickens. The weight of wing bones is much less than those of the wild ancestor, the jungle fowl, in proportion to weight of leg bones. This is specially true of the heavier breeds, since the Cochin and Brahma, for instance, very seldom use their wings.

Crossing.—Probably the most fruitful source of variation, and therefore evolution, is the crossing of different breeds or varieties. Before methodical selection was practiced, little consideration would be given to keeping breeds and varieties separate; and crossing was no doubt freely resorted to. Crossing adds to size and vigor, produces variations and abnormalities, restores lost or latent characteristics and increases fertility. Our present types and races of chickens were undoubtedly evolved in part from crossing.

Summary.—It is known, therefore, that all breed improvement is founded on variation. It is further known that variation may be induced by certain other conditions or factors. A change from one climate to another is a fruitful source of variation. This is true of plants as well as animals. A change of climate often gives increased vigor and fertility. Changes in climate and changes in food have undoubtedly had a great deal to do with the evolution of the fowl. Transplanted to a cold climate, we find the jungle type of three pounds has evolved into the twelve-pound fowl, because the fleshy fat fowl was better fitted to withstand the cold. On the contrary, the southern climates are not favorable to the heavy fowl, with heavy feathering and abundant fat, and as a consequence there was gradually evolved the Leghorn and fowls of that type.

Again, abundant food that was assured with domestication, undoubtedly exercised a potent influence in determining the size and characteristics of the fowl. Plenty of food tends to increase the size; scanty nutrition results in small races. Good food increases fecundity.

Crossing, however, is probably the most powerful means of variation. Crossing different breeds or varieties opens the door to further improvement, and to other breeds.

CHAPTER III

MODERN DEVELOPMENT OF INDUSTRY

Poultry-keeping had its birth, as has been seen, when the wild fowl of the jungle chose to foresake the wild way and become the companion of men. The domestication of the fowl and the beginning of poultry-keeping has been of tremendous importance to mankind. Poultry and eggs are more highly prized than any other form of animal food. The domesticated fowls are now producing in the United States over $600,000,000 worth of eggs and poultry annually, and the combined value of all poultry products of the different nations must reach a total of several billion dollars a year. Add to this the fact that the production and consumption of eggs and poultry are rapidly increasing, and a conception may be formed as to the magnitude of the fact of domestication. The development is not altogether a modern achievement. Men of modern times seem more concerned in exterminating wild game and animals than in preserving or domesticating them, and only the strong arm of the law has saved from utter annihilation many species of wild fowl.

Great as have been the achievements in the poultry realm under domestication, only within comparatively recent times has keeping poultry come to be recognized as an industry. Fifty years ago there was little or no poultry literature. The first enduring poultry journal was published in 1872 by H. H. Stoddard. Now poultry books are numbered by the score, and of poultry journals there are now half a hundred in the United States devoted exclusively to this industry, not to mention the mass of poultry literature published by the various farm journals as well as newspapers.

The first reference to poultry in the publications of the Department of Agriculture, is in the Annual Report of the Patent Office for 1845. The first bulletin or entire publication on this subject issued by the Department was Bulletin 41—"Standard Varieties of Chickens." Up to July, 1911, 41 publications devoted entirely to poultry, and containing 1,696 pages, were published by the Department. The Maine Station Report for 1887 was probably the first to report poultry work. The New York Geneva Station, Bulletin 29, 1891, was the first bulletin to report experiments with fowls.

It was in 1880 that poultry-keeping assumed sufficient importance to be included in a census of farm products by the federal government. Now practically every agricultural college and experiment station either has an organized poultry department or is giving instruction and conducting experiments with poultry in connection with other departments.

Notwithstanding all this recent work in poultry husbandry, history is strangely silent about the improvement of the hen or the development of the poultry industry, in which she is the most significant factor. Moving picture films came too late to tell her history. The word "fowl" or "cock" or "hen" is mentioned in books here and there throughout the centuries. As civilization advanced and books became more plentiful, more extended references are found, showing that the fowl was gradually coming into her own, becoming a factor of importance to civilized man.

Though there is no written history of the poultry industry until recent times, yet throughout the centuries the hen has been the companion of man, developing new characteristics, changing the color and pattern of her dress, adding to or subtracting from her weight, improving her economic qualities—all without a printing press, an experiment station, or a poultry show. The real history of the fowl preceded the poultry show and the poultry book. The improvement of the fowl was not all a matter of modern times. A century ago there existed all the sizes, large and small, that we have to-day, and if we are to believe some writers, economic qualities were as highly developed then as now. About seven hundred years ago eggs were so plentiful in Europe that they sold at the rate of 50 for one-quarter cent. At that time Charlemagne kept fowls on his "model" farms, and he himself prescribed methods of management of the fowls. The thirty-years war destroyed the poultry industry, as it did other industries; and there are those who maintain that at that time the secret of selecting the productive hen was lost and has never been recovered. Be that as it may, the care that was evidently given the fowls, and the cheapness of the eggs, might indicate that the fowls were very productive.

As an industry, the public is interested only in the economic aspect of poultry-keeping. The great increase in production has already been noted. This increase is probably without a parallel in the history of food production. What factors have been responsible for this increase? First, the increase must be ascribed largely to natural causes. With the increase in the percentage of the population that live in cities there has been a relatively greater consumption of eggs than of meat. There has been a greater call for a lighter diet than when the larger proportion of the population lived by toil or muscular labor. It can hardly be said that the increased price of meats has driven the people to eggs as a substitute, for the price of eggs twenty years ago was as low as 6 and 8 cents a dozen in different sections of the country, and yet the consumption of eggs per capita was less than it is now. With greater riches and higher compensation, the customers have turned more to eggs than to some other staple foods.

Second, a better knowledge of the high nutritive quality of the egg and of the fact that it cannot be adulterated—that it comes to the table in its original unbroken package, guaranteeing its purity—has also contributed to its increased use.

Third, cold storage, which is discussed in a later chapter, has also been a powerful factor in the increased use of eggs.

All this, by opening up larger markets for poultry products, has contributed to increased production, for without profitable markets no artificial stimulus could maintain increased production.

Education.—On the other hand, what may be called artificial stimulus was necessary. The demand for eggs would not have been fully met had education or artificial means not been resorted to in order to stimulate production. Under this head may be mentioned the agricultural and poultry journals. These journals have constituted a medium for an exchange of views by producers. Experiences have been published and re-published, and they have shown that there is money in producing eggs. Successes have been chronicled, and this has been followed by explanations of methods. In this way a great educational campaign has been going on through the medium of the agricultural and poultry journals.

The poultry page of the farm paper chronicling the poultry experience of successful farmers throughout the country, read by thousands of farmers weekly, has been a great force in directing attention to this industry and encouraging it among the farmers. It is doubtful if any other page of the farm paper has a larger circle of readers than the poultry page. The journals devoted exclusively to poultry, though they do not reach as large a constituency as the farm papers, have exercised great influence for better poultry. They have appealed more to the special poultry keeper than to the farmer, yet they have led in the dissemination of information along special lines, and thousands of readers have been kept informed by the poultry papers as to the progress being made in the industry. While making special appeals to the fancier or breeder of standard-bred poultry, these publications have paid more or less attention to the productive side of the industry, and they show a growing tendency to emphasize this. The general newspapers, both dailies and weeklies, are devoting an increasing amount of space to the campaign of poultry education.

The poultry show has also been an important factor in this development. It has afforded an opportunity for a study of breeds and external characteristics, and created an interest in the industry; this, too, in spite of the fact that the poultry show has been notoriously weighted down by standards of judging that in some respects handicap rather than encourage practical poultry breeding.

Though coming into the field late, the experiment station and the agricultural college have been rendering valuable assistance through state and federal aid. The results of investigations during the past fifteen years have been of distinct service to the industry; so has the work of the college in the teaching of students, and in institute work or extension work. The demonstration trains, in which the railroads co-operated with the colleges, have been the most successful agency in getting the information directly to the people interested. Moving picture films, industrial poultry contests among the school children and laying contests, are other agencies that are helping in the work of poultry development.

CHAPTER IV

CLASSIFICATION OF BREEDS

Breeds and varieties of fowls will be discussed here briefly from a utility standpoint. Before the days of the poultry shows and poultry books there were different races and breeds of fowls. There were the Mediterranean or Italian fowls, which were small of size, light feathering, active and nervous; and there were the Asiatics which were large, fleshy, heavily feathered and slow. These characteristics had been fixed before the business of breed making by the modern fancier had begun. The original Cochin weight has not been set any higher, and the minimum weight of the Italians has not been reduced. It would be difficult to conceive of any reason why there should be heavier breeds than the Cochin or Brahma, or lighter ones than the Leghorn, and yet if prizes were offered in poultry shows, or other rewards given for the largest fowls, it is without question that there would in time be evolved breeds of fowls of much greater weight.

From the jungle fowl, as we have seen, were evolved through the centuries the Asiatics of large size, and the Mediterraneans of small size. From these two pure races a hundred different breeds and varieties have arisen within less than a century. "The American Standard of Perfection" recognizes 121 breeds and varieties ranging in size from 12 pounds, to about three pounds, not counting the Bantams.

Standard Classification.—"The Standard of Perfection" classifies fowls according to external points of size, shape and color. It divides them into classes, breeds, and varieties. The class refers to the place of origin, the breed mainly to size and shape, and the variety to color within the breed. A full description of each breed and variety is given in an illustrated book called the "Standard of Perfection." This gives all the various exhibition points which go to make up the perfect specimen from the standpoint of the "Standard."

STANDARD CLASSIFICATION

Class	Breed	Variety
American	Plymouth Rock	Barred, white, buff, silver pencilled, partridge, and Columbian.
	Wyandotte	Silver, golden, white, buff, black, partridge, silver pencilled, and Columbian.
	Java	Black and mottled.
	Dominique	Rose comb.
	Rhode Island Red	Single comb and rose comb.
	Buckeye	Pea comb.
Asiatic	Brahma	Light and dark.
	Cochin	Buff, partridge, white, and black.
	Langshan	Black and white.
Mediterranean	Leghorn	Single-comb brown, rose-comb brown, single-comb white rose-comb white, single-comb buff, rose-comb buff, single-comb black, and silver.
	Minorca	Single-comb black, rose-comb black, and single-comb white.
	Spanish	White-faced black.
	Blue Andalusian	
	Ancona	
English	Dorking	White, silver grey, and colored.
	Redcap	Rose comb.
	Orpington	Single-comb buff, single-comb black, and single-comb white.

Class	Breed	Variety
Polish	Polish	White-crested black, bearded golden, bearded silver, bearded white, buff laced, non-bearded golden, non-bearded silver, and non-bearded white.
Hamburg	Hamburg	Golden spangled, silver spangled, golden pencilled, silver pencilled, white, and black.
French	Houdan	Mottled.
	Crevecœur	Black.
	La Fleche	Black.
Game and Game Bantam	Game	Black-breasted red, brown-red, golden duckwing, silver duckwing, birchen, red pyle, white, and black.
	Game Bantam	Black-breasted red, brown-red, golden duckwing, silver duckwing, birchen, red pyle, white, and black.
Oriental	Cornish	Dark, white, and white-laced reds.
	Sumatra	Black.
	Malay	Black-breasted red.
	Malay Bantam	Black-breasted red.
Oriental Bantam	Sebright	Golden and silver.
	Rose comb	White and black.
	Botted	White.
	Brahma	Light and dark.
	Cochin	Buff, partridge, white, and black.
	Japanese	Black-tailed, white and black.
	Polish	Bearded white, buff-laced, and non-bearded.
Miscellaneous	Silkie	White.
	Sultan	White.
	Frizzle	Any color.

Economic Qualities of Breeds.—It is difficult to classify breeds and varieties of fowls by their utility or economic qualities. When it comes to practical qualities it should be understood that no hard and fast classification can be given, because so far as egg production is concerned there is no known type or shape of fowl that indicates laying qualities with any measure of certainty. It has been demonstrated that there is a wide range in productiveness of fowls. Individuals of the same breed vary from no eggs to as many as three hundred in a year. High egg-laying is not a fixed breed characteristic; there are good and poor layers in all breeds. It is a question of individuals rather than of breeds.

And yet it may be conceded that high egg production is more often found in fowls of small size and active nervous temperament than in larger, less active kinds. The ability to go in the horse is usually associated with high energy and spareness of flesh. The cow that does things in milk production is spare in flesh, small in bone, and nervous in disposition. The little Shetland pony, it is said, produces power cheaper than the Clydesdale or Percheron It seems to require less fuel or food in the small animal to produce a given result, whether the result be milk, eggs, speed or power, than in the large animal. As the size is increased the cost of maintenance is increased. The large fowl is not the most economical producer of eggs. Frequently the large fowl will lay more eggs than the small fowl.

There is however, a relationship between size of fowl and egg-laying. But it cannot be said that high egg-laying is a fixed characteristic of any breed, or that there is any type that indicates with any certainty the laying qualities of the fowls when it comes to a question of selecting the good from the poor in any flock of any breed.

The system of trapnesting has shown us how widely individuals of the same breed and same type vary. There are "star-boarders" in every flock, birds that live on the thrift of others, possessing all the apparent external characteristics of their breed, but lacking the ability to lay. Whatever wonders modern breeders may have accomplished in the making of new breeds they have not given us a clear definition of egg type. So much has been accomplished by the fancier in the way of color breeding during the past fifty years that one is led to wonder what might have been performed if breeders had as persistently and intelligently bred for an egg type as they have for color types.

Practical Utility Classification.—All breeds and varieties of fowls may be grouped in four classes: 1, Egg Breeds. 2, Meat Breeds. 3, General Purpose Breeds. 4, Fancy Breeds.

1. **Egg Breeds.**—The most noteworthy characteristics of the egg breeds are: Small size, active and nervous temperament, early maturity, non-broodiness, good foraging habits, and sensitiveness to cold. The principal representatives of the egg breeds are Leghorn, Minorca, Spanish, Andalusian and Hamburg. All except the Hamburg, belong to the Mediterranean class.

2. **Meat breeds.**—Among the characteristics of the meat breeds may be mentioned large size, gentleness in disposition, slowness in movement, poor foraging proclivities, as a rule poor laying qualities, late maturity and persistent broodiness. Brahmas, Cochins, and Langshans are the principal meat breeds.

3. **General-purpose breeds.**—These are of medium size, are good table fowls, fair layers, less active than the egg breeds, but more so than the meat breeds, and are good sitters and mothers. Plymouth Rocks, Wyandottes, and Rhode Island Reds belong to this class.

4. **Fancy breeds.**—Bantams of various varieties; Polish

CLASSIFICATION OF BREEDS

and Silkies come under this head, and are raised chiefly for some peculiarity of form or feather without regard to useful qualities. This class will be eliminated from further discussion.

This classification, however, is an arbitrary one. Some breeders may object to the place given some of the breeds. It may, for instance, be claimed that the Langshan is as much a general purpose breed as the Orpington, and it may be that the Orpington is a better meat breed than the Langshan. The Orpington has a slightly greater weight than the Langshan, and if weight alone were to be considered these two breeds might exchange places. In making the classification, account is taken of the fact that the Langshan is largely, if not wholly, of Asiatic origin, while in the making of the Orpington several egg breeds were used. Again, placing the Orpington in the general purpose class does not mean that its meat qualities are not equal or superior to some of those in the meat class. The Dorking also, might fairly be placed among the table breeds because its table qualities have probably been more highly developed than its laying qualities, but on account of its medium size and its wide reputation as a general-purpose fowl it has been placed in the general-purpose class.

The classification includes only those breeds and varieties that have been admitted to the "American Standard of Perfection." There are many European breeds that are not illustrated in or recognized by the "American Standard." Some of them are of considerable economic value.

CHAPTER V

ORIGIN AND DESCRIPTION OF BREEDS

IMPORTANT EGG BREEDS

The Leghorn.—The poultry industry owes a great deal to the Leghorn fowl. It is not a made breed, as breeds are made to-day; it was "ready-made." Where the breed originated nobody knows. It is not the product of scientific breeding, but rather its type and characteristics have been developed through the centuries by the slow process of natural or unconscious selection. Nature early decreed that the high producer, whether the product be eggs or milk or speed, must be small in body, spare in flesh, and full of nervous energy. Nature did the work in the case of the Leghorn. While the Leghorn is a ready-made breed, our modern breeders have by careful selection given it greater uniformity, especially in color of plumage, ear lobe, etc. The development of different varieties has been the work of modern fanciers. Later and more productive strains have been developed, but the Leghorn of to-day is largely the Leghorn in type and characteristics of a century or two ago. Of all breeds of fowls, few have the apparent lasting qualities of the Leghorn. While the Leghorn is a large class at all poultry shows, and has therefore been bred along fancy lines, it has also been bred for special egg-laying qualities. The White Leghorn has the distinction of being found on special poultry farms more than any other breed.

ORIGIN.—The Leghorn is sometimes spoken of as an American breed. It received its name in the United States, but the fowl came from Italy and derived its name from the city of Leghorn, Italy. In Italy and other European countries it goes under the name of Italian fowl. As a fancier's fowl it may fairly be said to be an American production, its finer exhibition points being put on by American breeders; but its general breed characteristics were developed in the Mediterranean country before the fancier himself was developed in America. In Italy not so much attention has been given to color, Alfredo Vitale, of Naples, in a letter to the writer, expresses the opinion that the blacks are the most productive strain. It is claimed that the Black-Red Game was crossed with the Brown Leghorn to improve the color of plumage, also that the Buff Cochin blood was used to secure the proper buff color in the Buffs.

WHITE LEGHORN COCKEREL
Exhibition type.

There is little difference in laying in different varieties of Leghorns. A mixture of Cochin blood in the Buffs and Game blood in the Browns may have had an influence toward lower egg yield, but it would hardly seem probable that the effect of that infusion of blood from less productive breeds would still remain. It was ill-advised, however, to jeopardize well-known laying qualities by crossing with meat breeds because of a color demand.

The Minorca.—Among the egg breeds, next to the Leghorn, the Minorca ranks in popularity. Like the Leghorn, its type was fixed long ago. It is larger in size than the

Leghorn and has light colored skin and dark shanks. Its strong point is that it lays a large white egg. No other breed of fowls lays an egg as large and attractive as the Minorca. In markets where the white egg is preferred Minorca eggs should command the highest price. The Leghorn excells the Minorca in number of eggs, but it may be possible to secure a price for the Minorca eggs so much higher as to make the product of the latter equal that of the former in value. Minorca eggs frequently weigh as much as 28 ounces a dozen, and a good average would be 26 ounces. Large size of egg is characteristic of the Spanish breeds. Under proper conditions, the Minorca is an excellent breed to keep. In the southern or warmer sections of the country it thrives, but its excessively large comb and wattles make it hardly desirable for the cold sections. With proper shelter, however, it will do well. The Minorca derives its name from the Island of Minorca off the east coast of Spain. There are two varieties, Black and White. "The Standard" subdivides the Blacks into single and rose comb varieties. The characteristics of white skin and dark shanks depreciate their value somewhat in American markets.

BLACK MINORCA MALE

The Ancona.—During the past few years the Anconas have been receiving considerable attention. They have mottled white and black plumage. This fowl is undoubtedly of Italian origin, and outside of the distinguishing feature of color it is pretty much a Leghorn in type and characteristics. Compared with the Leghorn it is comparatively rare in this country, and its egg-producing qualities have not been so well demonstrated.

The Hamburg.—"The American Standard of Perfection" classes the Hamburg as a Dutch breed, while Edward Brown classes it among the British races of fowls. There are six varieties, namely, Golden and Silver Spangled, Golden and Silver Pencilled, White and Black. The Hamburgs all have rose combs. They lay a rather small egg, though the Blacks, owing probably to an infusion of Spanish blood, lay a fair size egg.

The Blue Andalusian.—This is another of the Spanish egg breeds. The fowls have a considerable popularity as egg layers, and lay an egg of fair size. A peculiarity of this breed is that though blue is the recognized color the mating of two blues together produces offspring that are either black or splashed white. In mating the blacks and whites together usually blue offspring results.

The Black Spanish.—This was a very popular breed thirty years ago. The fowls were splendid layers of a large white egg, but are now very seldom found in any section of the country. The breed has fallen the victim of a too general tendency for fanciers to accentuate in their breeding special points or peculiarities. In this case the peculiarity was the long white face; breeders engaged in a rivalry to increase its length. This was encouraged by the "Standard" which says of the white face: "The greater the extent of surface the better." The Black Spanish is of the same family as the Minorca.

The Campine.—This is the most popular egg breed in Belgium. We give a description of two Belgium breeds for the reason that in that country, as might be expected from the character of the people, fancy characteristics have been given slight consideration in the breeding of fowls. The frugality and thrift of the Belgian peasants would lead one to expect that they would exercise great care in the selection of breeding fowls, and the chances are that these very characteristics make them more skillful in selecting the best for breeding. This breed is of great antiquity, and as the Belgians believe strongly in the egg basket the seeking after abnormalities or fancy points was not permitted to deter them from their pursuit of the egg-layer. It is a non-sitting breed, an excellent layer of white eggs of good size. In size it is about the same as the Leghorn; it has dark slate blue legs and feet. There are two varieties, the Gold and Silver.

The Braekel.—This is another breed that has been bred a long time in certain districts of Belgium. Its origin is probably the same as that of the Campine and any differences now in size and characteristics are probably due to differences in environment. It is larger than the Campine, females weighing from 4 to 6 pounds, and males 5 to 7 pounds. It is very precocious. One writer says: "The chicken is no sooner out of the shell than its comb is developed; at three weeks the cockerels commence to crow; at six weeks they begin to drive about the hens." The Braekel, it is stated, is as much developed at six weeks as some other breeds at about six months.

The Houdan.—In France more attention has been given to developing meat qualities in fowls than laying qualities. In this line the French poultry keepers are particularly apt. Much may be learned from the poultry raisers of France in the production of a fine quality of table meat. Mr. Edward Brown in his classification of fowls places only one French breed—the Houdan—among the egg breeds.

Even the Houdan may properly belong to the general purpose class on account of its weight and meat qualities. The weight of the adult female is 6 pounds. The crest is more ornamental than useful and this peculiarity has undoubtedly had a great deal to do with the lack of appreciation shown for it among utility poultry-keepers. It lays a white egg, the color of legs is pinky-white, mottled with black.

GENERAL PURPOSE BREEDS

American breed makers have run altogether to the general purpose type of fowl. In the American class we have the Plymouth Rock, the Wyandotte, Java, Dominique, Rhode Island Red, Buckeye, all of medium size and of general purpose characteristics. Out of those six breeds twenty varieties have been made, there being of the Wyandotte alone eight different varieties, the differences in varieties being wholly in color. It is true that American breeders have "made over" other breeds that have come from foreign countries until some of them would scarcely be recognized as of the same breed, but the work has been chiefly confined to fixing external points of color, not in altering type.

The general-purpose fowl is a modern innovation. Before the days of the modern breeder there were practically but two types of fowls—the large, slow, fleshy Asiatic, and the small egg-laying Italian. The modern breeder has concerned himself not so much in improving these two types as in making various combinations of them. Our American breeds are therefore the result of crossing the two pure races mentioned. The poultry industry has doubtless gained from the making of these varieties. A question naturally arises as to whether it would not have been better if the breeders had confined themselves to improving existing breeds or races and keeping them pure rather than mixing them and making new breeds.

If it be true that crossing improves the vigor and fertility of the offspring and saves races from annihilation, it may readily be conceded that the amalgamation of the two races for the purpose of making new breeds has been altogether an advantage. If it be denied that the crossing of pure races can ever be beneficial we will have to confess that our American breeds and varieties are without excuse of origin. It is certain that the Italian or Leghorn has not been replaced by a better laying breed, nor has there been produced a breed superior to the Cochin and Brahma as meat breeds. In American breeds the excessive weight of the Cochin and Brahma has been avoided and the preference of the largest proportion of consumers for a medium size table fowl has influenced American breeders in the making of new breeds, to the undoubted advantage of the industry as a whole.

When we speak of a general-purpose breed reference is made to meat and egg-laying qualities. In other words, a general-purpose fowl is a fair layer and a fair table fowl, and that idea has been kept in mind by the originators of the Plymouth Rock and other breeds of that type. They wisely eliminate feathers on legs, which are objectionable in a utility fowl. They also eliminated some of the natural wildness of the Leghorn. Breeders have been somewhat hampered by the demands of the show which required them to select for various other points, and for this reason our Plymouth Rock and other general-purpose breeds have not been bred up to that perfection of flesh that has been attained in some of the French and in some of the English breeds. Undoubtedly the craze for fine barring in the Barred Plymouth Rock has engaged the attention of many breeders to the exclusion of points demanded in a good

ORIGIN AND DESCRIPTION OF BREEDS

table fowl. As a result there is more uniformity in the barring, for example, than in the proportion of edible meat to bone in the Plymouth Rock. As a rule the Plymouth Rock is too heavily boned for a fine table fowl.

Another objection to the Plymouth Rock and other general-purpose breed may be urged; the "Standard" weight is larger than is demanded by the great body of consumers. The general-purpose fowl should fill a general purpose demand. "The Standard" weight for the Plymouth Rock is 9½ pounds for the cock and 7½ for the hen. If the judge must not cut for over-size they are placed practically in the Asiatic class so far as size is concerned. And as the size is increased the breed is getting that much away from the general purpose type. Increasing the size does not necessarily mean better meat qualities.

A fowl that exceeds 7 or 8 pounds in weight borders too closely on the Asiatic or meat type for a general-purpose fowl. When it reaches 8 or 10 pounds it gets into a special class and there must be a special market for it. The demand is limited for the large meat type of fowl, and if the poultry raiser is to meet the requirements of the largest body of consumers he must breed a fowl of medium size. It would be an economic mistake to advocate a large fowl of Asiatic type for the general farmer, because if all were to adopt that type it would mean one of two things: The cutting of the consumption of poultry in two, or cutting the price in half. There is a greater demand for a fowl weighing 4 to 5 pounds dressed than for one of any other size. The problem then for the breeders of a general-purpose fowl is to adhere to a type that will meet the largest consuming demand, and then develop laying qualities on that basis.

Again, breeders of the Plymouth Rock might render a real service if they should eliminate the tendency of the

Plymouth Rock to put on excessive abdominal fat. However, this result will largely be secured in breeding for eggs. It will be found that this characteristic is usually absent in the heavy producer.

In our American breeds many of the Asiatic characteristics are retained. In some respects American ideals differ from European. We get the brown egg from the Asiatics, but this is one of the accidents of choice, for American markets generally prefer the white egg shell. On the other hand we get the yellow leg and skin from the same source, the color preferred in the markets. In England and other European countries, the white skin, it is believed, indicates superior excellence of meat, but with this white flesh they get something they don't want—a white egg. Here are apparently antagonistic characteristics, a white egg and a yellow skin in general purpose breeds. Whether it is possible to overcome this barrier of nature remains to be demonstrated.

Speaking of American breeds, Edward Brown pays a high compliment and at the same time extends a warning to American breeders in the following language: "That these breeds have proved most valuable additions to our stock is unquestionable, and their wide distribution and universal recognition is a great tribute to American breeders, who have kept prominently forward the general economic qualities and not exaggerated special points to the extent met with in Great Britain. Whether that will be so in the future remains to be seen, for present signs are in the direction of an exaltation of fancy points, which would be regrettable."

If we are to judge from the success of one English breeder in American laying competitions, the warning has been better heeded in England than in America.

ORIGIN AND DESCRIPTION OF BREEDS

The Plymouth Rocks.—There are five varieties of Plymouth Rocks recognized by the "American Standard of Perfection"; namely, the Barred, White, Buff, Partridge, and Columbian.

The Barred Plymouth Rock is the great American farm fowl. Its popularity among farmers exceeds that of any other one breed. The White Rock, the Wyandotte and the Rhode Island Red may be of equal utility value, but the Barred Rock has been longer established, is more widely known, and its qualities of meat and egg production, and possibly its color, have given it a place second to none. The Barred Plymouth Rock was about the first American production in the poultry world, and on this account it no doubt secured a popularity that later productions did not. To the fancier fine barring in the Plymouth Rock represents the highest achievement in breeding. When associated with this is good shape and carriage, the Barred Plymouth Rock is a most attractive fowl.

BARRED PLYMOUTH ROCK COCKEREL
First prize at Los Angeles show.

As to the origin of the Plymouth Rock, several Asiatic and Mediterranean breeds are represented among its ancestors. It is believed that the first or original cross was a mating of Dominique male and Black Java or Black

Cochin hen. The Minorca, Cochin and Brahma are all believed to have been used in making the breed. Mr. D. A. Upham and Mr. Joseph Spaulding, both of Connecticut, claim the honor of originating the Barred Rock. The former exhibited the first specimens in 1869. Close breeding for fine barring has injured some strains of the breed, but its wide dissemination has averted ruin. A peculiarity of the Barred Rock color is that the male offspring are lighter than the female. The tendency is for the cockerels to be lighter than the parents, and the pullets darker. The "Standard" says that show specimens must have the same shade of color, male and female, and this has led fanciers to follow in breeding what is called double mating.

BARRED ROCK HEN
Showing fine barring. (Courtesy of Miller Purvis.)

Double mating means the use of two separate matings, or two separate pens of fowls, one to produce males of proper exhibition color, the other females. The pen producing the cockerels is darker in color than the pen producing the pullets.

The White Plymouth Rocks.—There is no difference in the shape and size of the different varieties of Plymouth Rocks. The difference is in color only. The White came as a sport, or what might be called a "mutant" from the

Barred variety about 1880. The White is fully the equal of the Barred in economic qualities. It has not been subjected to the same degree of intense inbreeding as many strains of the Barred Rock have been for barring, and on that account there may be excuse for the claim that it has better maintained its original utility value. However, there are strains of the Barred variety that by good breeding have preserved their utility value to a high degree.

The Buff Plymouth Rock.—This variety of Rock is an independent creation. It is not related to the Barred or White varieties, but breeders have so moulded it and shaped it that in size and type and general external characteristics it is a duplicate of the others. The Buff Leghorn, Buff Cochin and Light Brahma were used in producing the Buff Rock. Some strains, it is claimed, have originated from the Rhode Island Red. While it is a fowl of much merit, the Buff Rock is not a popular breed on the general farms, or on special poultry farms, and it has never been demonstrated that it has any useful qualities not possessed by the Barred or White variety. From an economic point of view there can be little excuse for the Buff variety.

Columbian Plymouth Rock and Partridge Plymouth Rock.—These newer varieties of the breed, vary only in plumage color. They have been established, as other varieties have been, by a system of cross and inbreeding and admitted to the "Standard" as a new breed because they have a distinctive color, not because of any difference in real practical value.

The Wyandottes.—The Wyandotte was the second production of American breeders. As a breed, the Wyandotte has a type of its own. In size and shape it meets the requirements of a general-purpose fowl probably better than any other American breed. In size it is a little smaller than the

Rock, and of a more blocky build. Its blocky shape and comparatively early maturity make it a good broiler breed. The originators, therefore, had a valid excuse for giving it to the world as a new breed. An objection may be urged against the Wyandotte that its type is not very firmly fixed. Wyandottes are frequently found of the type of Plymouth Rock or Rhode Island Red rather than of the blocky, compact build. While they are fully the equal of the Plymouth Rocks in egg production, their eggs average smaller in size. These are points that the breeders may rectify. If it should develop that high egg production is found in fowls of the long body or rangy type, then the Wyandotte would have little excuse for existence as a general-purpose utility fowl. However, it has not been proved that a long body is a sure indication of good laying qualities.

WHITE WYANDOTTE HEN
(Courtesy of A. G. Duston, Massachusetts.)

As to origin, the Wyandotte came by accident rather than by design. If some authorities are right, a Sebright Bantam and a Cochin hen were mated together to produce an improved Cochin Bantam. Silver Spangled Hamburg blood was added; then another cross and a half-bred Cochin hen was used. The breed was given the name of the American Sebrights, later the Wyandottes. The first Wyandottes were produced in the 70's.

ORIGIN AND DESCRIPTION OF BREEDS

Variety makers have found a fruitful field in the Wyandotte family. There are now eight different varieties of Wyandottes, as follows: Silver, which was the first, Golden, White, Black, Partridge, Silver Pencilled, Buff and Columbian.

The Rhode Island Reds.—This is one of the most popular breeds in America. More than any other American breed it owes its distinction to its practical qualities. For many years it had been bred as a farm fowl, establishing a reputation for real merit, before it was taken up by "Standard" breeders and admitted to the "Standard" as a

WHITE WYANDOTTE COCK
(Courtesy of W. D. Kelley, Oregon.)

breed. In weight it is the same as the Wyandotte, but in type it shows more of the Plymouth Rock characteristics than the Wyandotte. It is less suggestive of the Cochin than the Wyandotte. That it is a fowl of great merit is attested by the fact that in the Little Compton poultry district of Rhode Island, where the poultry industry has been developed to a larger extent than in any other district of the continent, with the possible exception of Petaluma, Cal., it is almost universally kept on the farms. The Rhode Island Red has been in existence possibly more than a century, but fanciers were slow in taking hold of it, and not until a few years ago was it admitted to the "Standard of Perfection." It is now a prominent class at all poultry shows of the country. This breed originated on the farms as a practical fowl and with little or no thought of making a new breed; that it has gained so great a popularity is proof that the poultry-keepers are alive to the importance of utility qualities.

Considerable obscurity naturally attaches to its origin. It is believed, however, that its foundation was the common farm fowl on which were crossed breeds of Asiatic

RHODE ISLAND REDS
(Courtesy, Howard H. Keim, Oregon.)

as well as Mediterranean blood. There is only one variety of Reds, though there are both single and rose comb strains.

McGrew says that this breed is: "The result of fifty years of careful outbreeding, and it would have been better for the stamina of many of our breeds if they had been bred on the same plan, instead of inbred." To Dr. N. B. Aldrich of Fall River, Mass., is generally given the credit of introducing the Rhode Island red to the public as a new breed.

The following interesting facts as to the making of this breed are given by Miller Purvis: "All over the country, men who had sailed the seas, brought home fowls from India, China and Europe. These fowls were crossed and mixed in indiscriminate confusion. Red Malay, Shanghai, Chittagong, Brahma and Leghorn were bred and crossed in every conceivable way." The idea of making a new breed finally came to Dr. Aldrich and Mr. Buffington. "They did not agree on the exact shade the bird should be, and each selected those which suited his fancy. Mr. Buffington called his Buff Plymouth Rocks and Dr. Aldrich invented the name of Rhode Island Reds for his, and each took hold of the public fancy and two new breeds were born from the same flock, both of them being of mongrel blood pure and simple."

The Dorking.—The Dorking has frequently been spoken of as the grand old breed of Great Britain. It is an ancient breed, attaining popularity long before the introduction of the Cochin or Brahma into Europe or America. So ancient is it that some enthusiatic writer has said of it: "It would be vain to attempt to trace the origin of a breed which was accurately described two thousand years ago by a Roman writer; and as Roman stations abound in Cumberland it is quite possible that a poultry-fancying praetor fifteen hundred years ago might send or carry in the same year the first couple of Dorking fowls to the bank of the Thames."

Be that as it may, it is certain that Dorking is a breed of antiquity, as well as a breed of great merit for meat qualities. The Cochin craze of sixty years ago threatened its existence, but the English breeders stood to their guns and saved it from amalgamation with an inferior race. While a large proportion of our general-purpose and meat breeds have an infusion of Asiatic blood to the extent of dominance, the Dorking successfully weathered the craze, and is to-day as pure in blood apparently as though the Cochin had never been known. As evidence that it does not owe its size to the influence of Asiatic blood it is pointed out that there are records which show that the Dorking attained a weight of 14 pounds more than a hundred years ago, or fifty years before the introduction of the Cochin. Brown says that the Dorking "by its fineness of flesh and delicacy of skin, the whiteness of the flesh and legs and the abundance of meat carried upon the body, must be regarded as one of the best table fowls that it is possible to obtain." In spite of the fact that it lays a white egg it maintains much popularity in Britain. It cannot be said to be a heavy layer, though we are inclined to place it among the general-purpose breeds.

The Orpington.—The modern Orpington is now dividing honors in England with the Dorking as a general utility breed. From the English standpoint, in one respect at least, the Orpington has an advantage over the Dorking; that is color of egg. The combination of white legs and skin with tinted or brown eggs is the peculiar achievement of the Orpington makers, just as the Plymouth Rock laying a white egg would be an achievement for American breeders in meeting the market demands. This new combination is no doubt largely responsible for the popularity of the Orpington in Great Britain. The combination of white skin and brown egg, however, though commending it to the buying public of Great Britain, handicaps it as a competitor in America with general-purpose breeds. Brown places it among the general-purpose breeds, but from its weight and meat qualities it might well be placed among the meat breeds of this country. If our market preference for yellow legs and skin and white egg is to be maintained, it is difficult to see why our American general-purpose breeds should be replaced by the Orpington. If we wish to discard our American breeds it would be more consistent to take up the Dorking, which is fully the equal of the Orpington as a table fowl and in addition lays a white egg.

The Buff Orpington.—The origin is somewhat clouded in obscurity. Brown and other English authorities argue that the Buff Orpington came from a farm fowl known locally as the Lincolnshire buffs, and that its real origin was the Dorking crossed on the common fowl, intercrossing with Buff Cochin. William Cook, however, is usually given credit in this country for originating the breed, and he claims that he crossed the Golden Spangled Hamburg and the Buff Cochin, and then the Dorking. Brown states that there is abundant evidence that the great majority of the present-day Buffs are directly bred from Lincolnshire buffs without the slightest relationship to Mr. Cook's strain. Mr. R. de Courcy Peele says: "The foundation had been laid many years previous to Mr. Cook's time in the shape of the Lincolnshire buffs, a variety, if it may be so called, which has for many years been the acknowledged farmer's fowl in and about Spaulding and the neighboring towns."

The Black Orpington.—There seems to be no question that William Cook was the originator of the Black Orpington. The interesting point in its origin, according to Mr. Cook himself, is that its ancestors were rejected specimens of Black Minorca, Black Langshan, and Plymouth Rock (black). The Minorca had such marks as red earlobes; the Langshan no feathers on legs, and the Plymouth Rock fowls were black. This is a mongrel origin, so far as present exhibition points are concerned. From such an origin we have one of our most beautiful breeds of fowls and one of considerable utility.

The White Orpington.—The White Orpington is said to have been produced by a combination of White Leghorn,

Black Hamburg and White Dorking. Brown offers objection to this origin on the ground that as two of these breeds have rose combs, the White Orpington of the present day would show the rose comb very frequently. Later knowledge of breeding, however, teaches that if proper selection be made of the crossed offspring, the characteristic rose comb need not ever show in subsequent generations. Data on this point is given in Chapter III, page 79. Brown declares his belief that the White Orpington originated as a sport from the Blacks.

WHITE ORPINGTON HEN
A noted prize winner.

The recent popularity of the White Orpington in this country is striking evidence of the power of printers' ink. While the breed undoubtedly has great merit, there is no real reason why it should displace our American breeds which are not handicapped by white skin and legs. Unless a new breed can be shown to have superior egg-laying qualities it is a mistake to advocate it as a utility breed when it possesses other characteristics which depreciate its value as a market fowl. If certain breeders wish to cater to the fancy trade, well and good; if their effort is to produce something which will delight the eye and sell for fancy prices on that account, that should be clearly understood. But utility values should not be set by the amount of printers' ink used in advertising. Where the real standard is the market demand for meat and eggs, it is a self-evident business proposition that we should choose fowls possessing in greatest measure the characteristics demanded by the market. These suggestions apply to other breeds as well as to the White Orpington, but few breeds have been boomed as the latter has. It is a breed, however, of distinctive merit, but as a market fowl it fills the English rather than the American market demand.

The Sussex.— Some authorities would place the Sussex ahead of the Dorking as the grand old breed of England. It seems to be an equally ancient breed. Its chief point is its meat quality.

LIGHT SUSSEX
Owned by J. H. Barker, California.

Wright speaks of it as surpassing "every other breed on earth" in this respect. It has made the fattening industry of certain districts of England famous. It has something of the shape and type of the Dorking, but somewhat smaller; it has four toes, and lays a tinted egg. It is broad in back, full-breasted, fine boned, and hardy. Barring the defect of white skin, it is a type of fowl that might well be used in this country for a market fowl. There are three varieties: White, Speckled, and Red.

The Faverolle.—In studying general-purpose fowls account should be taken of the French breed, Faverolle. In

SPECKLED SUSSEX HEN
Owned by J. H. Barker, California

size it belongs to the general-purpose class. In quality of meat it is excellent, and is a fair layer. It has a large, deep, and broad body, rather short in legs and small in bone. Other characteristics are white earlobes, white skin, heavy beard and muffs, slightly feathered on legs. The fowls are very tame, and stand confinement well. In meat qualities they probably surpass our American breeds, but they have the same handicap as the Orpingtons—white skin and brown eggs.

IMPORTANT MEAT BREEDS

There are few breeds, and none of them have originated in America, that are specially made for meat production. Breeding for meat in this country is practically an un-

SPECKLED SUSSEX COCK
Owned by J. H. Barker, California.

ORIGIN AND DESCRIPTION OF BREEDS

used term. Very little earnest concerted attention has been given to breeding for excellence of table qualities. This will be a development of the future. Were there a standard of excellence that would disqualify a fowl, or throw it out of the market, that did not show at least 25 to 30% more meat than bone and offal at six months of age, there would soon be a change in the meat qualities of the fowls found in the markets.

FAVEROLLE HEN
(Courtesy, Editor "La Vie a La Champagne," Paris.)

DOMESTICATED

In a good table fowl there should be a large percentage of edible meat and a relatively small amount of bone and offal. Heavy bone and frame should not be developed at the expense of meat. Fowls vary greatly in this respect. Mons. E. Lemoine, of France, has published the results of some investigations on this point, as follows:

WEIGHTS OF MEAT AND BONE ON FOWLS OF DIFFERENT BREEDS [1]

	Weight of meat on fowl 6 mos. old			Weight of bone, etc., on fowl 6 mos. old		
	lbs.	oz.	grs.	lbs.	oz.	grs.
Barbezieux	4	10	92	4	15	0
Cochins, buff	4	9	0	5	4	327
Courtes Pattes	3	10	99	2	8	316
Crevecœurs	4	9	66	4	14	197
Dominiques	3	11	66	2	8	279
Dorkings, silver-gray	5	4	282	4	13	403
Du Mans	4	6	64	2	11	11
Game, brown-red	3	15	233	2	7	301
Hamburghs, pencilled	1	15	335	2	7	224
Hamburghs, spangled	2	3	236	2	7	301
Houdans	3	7	0	2	10	140
La Bresse, gray	3	7	67	2	8	163
La Bresse, black	3	7	375	2	8	240
La Fleche	3	5	339	2	9	269
Langshans	5	4	359	5	1	78
Leghorns	3	15	233	2	10	140
Polish, spangled	2	12	348	2	8	18

There were evidently inaccuracies in the work on which this table is based. The Leghorns are given a weight of over six pounds. They were not the Leghorns that we know to-day. It is evident, however, that breeds vary greatly in respect to the point under consideration. It would be instructive if the data could be extended to include our several American breeds.

The table on p. 54 gives further data on this subject from work at the Oregon station. The fowls used were the Barred Plymouth Rock and White Leghorn, and first crosses of those breeds:

[1] From "Poultry-Keeping as an Industry for Farmers and Cottagers," by Brown.

THREE BREEDS OF DIFFERENT TYPES

Representing heavy, medium and light breeds. Each made the highest egg record in its year in Australian Laying competitions. High egg production is more a question of breeding than of breeds, of heredity than of type.

Breed of Fowl	Plymouth Rock		B.R.&W.L. Cross		White Leghorn	
Number of fowl	74	90	269	276	85W	86♂
Live weight (lbs. oz.)	7-0	7-0	5-0	4-7.7	2-14.94	3-6.82
Picked and bled (lbs. oz.)	6-8	6-8	4-11	4-1.6	2-10	3-1
Per cent loss in picking and bleeding..	7.14	7.14	6.41	8.41	10.43	9.66
Weight after drawing (lbs. oz.)	4-12.9	5-0.7	3-11.6	3-1	1-14.2	2- .8
Per cent loss in drawing	26	22.42	20.39	25.32	27.99	31
Weight of head, bones, shanks (lbs. oz.) ..	0-8.2	0-7.9	0-6.6	0-6.4	0-3.9	0-3.9
Total weight of meat (lbs. oz.)........	4-4.7	4-8.8	3-5	2-10.6	1-10.4	1-12.9
Per cent of edible meat to live fowl	61.33	64.9	66.2	59.4	56.1	52.7

It is seen from the above that there is a larger percentage of edible meat in the Plymouth Rock than Leghorn. In this test there was 15% more. The cross-bred showed practically the same amount as the Plymouth Rock. This indicates an important difference in the meat value of different types of fowls.

Meat breeds should possess a finer quality of flesh than general-purpose breeds; but this may not always be evident in the breeds as we find them. Heavy egg production is not and probably never will be associated with excellence in meat quality. The active nervous disposition of the egg breeds is not favorable to the production of meat of high quality. Good meat quality, therefore, should be looked for among the slow, inactive, docile breeds. Again,

ORIGIN AND DESCRIPTION OF BREEDS 55

good meat quality will not be found in fowls of large bone and heavy feathering. In addition the meat breeds should have large size. The only breeds of any prominence in America that will come under this classification are the Cochin, Brahma and Langshan.

The Brahma.—There are two varieties of Brahma, the light and the dark. The light Brahma is the largest vari-

GOOD UTILITY BARRED PLYMOUTH ROCK BREEDING COCKERELS
(Oregon Station.)

ety of fowls of any breed. The Dark Brahma is a pound lighter according to "Standard" weights. The chief characteristics of the Brahma are: Large size, gentle disposition, slow, easily confined by low fences and long in maturing. The fowls lay a brown egg, have yellow skin and shanks, and have heavily feathered legs. There are good layers among them, but to breed them specially for laying they would most likely degenerate in their meat qualities. To maintain them as a meat breed, egg-laying should be secondary. The breeder should choose his breeding stock from those of good meat type and be satisfied with a fair yield. The special point of the Brahma is in filling a demand for large roasting chickens. In some markets there is a strong and growing demand for large roasting chickens, and the Brahma fills the demand pretty well. The Brahma as a utility fowl should fill an important place in the poultry industry, but in breeding it the market demands must be the only standard of excellence.

LIGHT BRAHMAS
(Courtesy of E. Shearer, Oregon.)

The origin of the Brahma has occasioned some controversy. It has been claimed that it was made in America, but this is disputed by the best authorities. There is no doubt of its Asiatic origin. Brown asserts that the original type of Brahma is met with in the Brahma-pootra district of India. The original Brahmas were light in color, the dark variety being the result of breeders' work in England and America. They were imported into the United States about 1846, and a few years later into England. The type has been changed considerably, more especially in England where the breeding of fancy feather points, especially leg feathering, has been carried to the extreme.

The Cochin.—The meeting of the yellow Asiatic race with the white race took place in 1846—speaking of races of poultry. Nothing disturbed the poultry world like the invasion of the yellow Cochin. It was lauded to the skies, just as it was bitterly execrated. Wright says: "It was averred that there was no property that a good fowl should have, but this possessed it; it was delicious roasted or boiled, and the hens laid two or three eggs a day." Again, he says: "Loud and long were the protests made by the best utility breeders, but these were written down by the glib pen of the ignorant but ready writer." He tells us further: "One of the greatest evils that befell the splendid, large, well-formed and profitable table fowls of the southern counties was the introduction of the Shanghai or Cochin." Again: "Then came the Shanghai fowls and the craze for size, novelty and colored eggs; and ill it fared with our old breeds." "The Cochin or Shanghai craze was the first blow that our ancient and almost perfect farm poultry received."

BUFF COCHIN HEN
Exhibition type of present day. (Courtesy of Dr. J. J. Hare, Ontario.)

Then he sarcastically says of the Cochin: "They were to furnish eggs for the breakfast, fowls for the table, and better morals than even Doctor Watts' hymns for the children, who were from them to learn kind and gentle manners and thence forward to live in peace." In 1847 it was declared that "all England was given over to a universal hen fever" —the Asiatic had invaded England, and the Britisher had bowed the knee. It is asserted that $250 was freely paid for a cock bird, and $25 for a sitting of eggs. The Queen of England had received an importation in 1845. Poultry

LE MANS
(A special French meat breed. (Courtesy of Editor, "La Vie a La Campagne," Paris.)

shows became fashionable and great crowds attended, but in process of time, as Brown says, "the bubble burst." Wright declared valiantly against the crime of "mongrelizing" the ancient fowls of England with the Cochin.

The invasion occurred. The Cochin disappeared, but not before the fowls of two continents had been "mongrelized" as Wright would have it. The Cochin has practically ceased to be a part of the poultry industry. It is unknown as a practical breed. It has passed from the stage. But it has left its stamp. The yellow leg and the yellow skin came from the Asiatic, so did the brown egg. All our prominent American-made breeds and many prominent European breeds have a mixture of Cochin blood. They are all tainted. Like the bee that stings and pays the penalty with its own life, the Cochin has suffered annihilation.

What the practical effect has been of the amalgamation of the two races, the yellow and the white, it would be diffi-

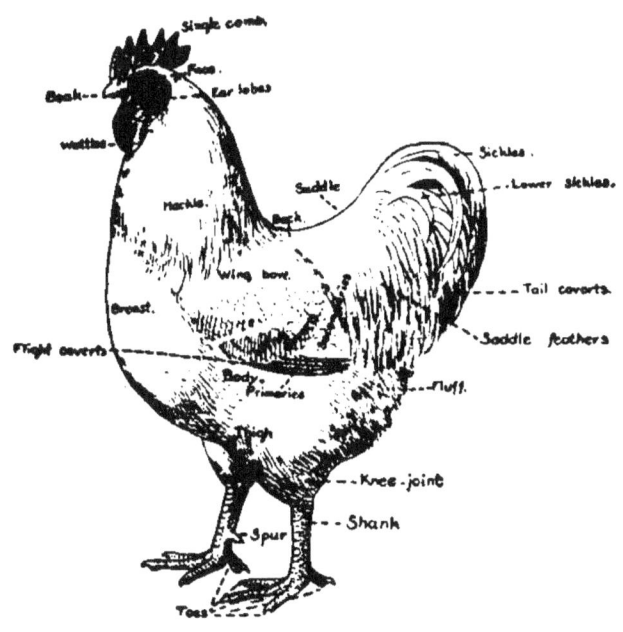

POINTS OF THE FOWL.

cult to say; but there will be few who will now take the stand that the poultry industry was badly stung by the Asiatics. As the case now stands, the Cochin blood is found in a great many breeds that excell the present-day Cochin in practical qualities. The extinction of the Cochin as a practical breed was due not so much to any glaring demerits of the fowl, but rather to a system of breeding for fancy feathering which made it impossible to maintain in the breed whatever useful qualities it may have possessed. The comparatively little emphasis placed on meat qualities by poultry-keepers in general was no doubt also a factor in its extinction.

There are four varieties of Cochins; namely, Buff, Partridge, White and Black. The chief characteristics of the Cochin are its large size—loose feathering giving it a more massive appearance, and gentle disposition. In shape it is short, b r o a d and blocky.

LA FLECHE
(Courtesy, Editor "La Vie a La Campagne," Paris.)

The Langshan.—The Langshan also originated in China, but like the Cochin and Brahma it has been improved in appearance by the fanciers. It has a greater popularity for laying than the Brahma and Cochin, but inferior as a meat fowl. A pen of Langshan fowls made a wonderful egg record at the Australian laying competitions, the particular pen being from stock imported direct from China, and represented a different type of fowl than the Langshan now found in this country.

The La Fleche.—The La Fleche is one of the leading fowls of France. High prices are paid in France for large fowls of good quality, and this breed is largely used to supply this demand. The males weigh up to ten pounds and the females to eight. The plumage color is black. In this breed the French poultry-keepers have evolved a fowl of great merit. Its flesh has exceptional delicacy.

CHAPTER VI

PRINCIPLES OF POULTRY BREEDING

The breeding of poultry for definite types or characteristics is a modern art. Even at this day, in its general practice, it is largely a hit or miss business. It cannot be said that it has been reduced to a science if by that is meant that the breeder can predict with accuracy the results of his work. It has been contended only within the present decade that, on the one hand, the hen does, and, on the other hand, does not transmit laying qualities. The preponderance of evidence seemed to favor the view that heredity counted for little or nothing in the science or art of poultry breeding as it related to improvement of egg-laying qualities. The stage had been reached, it was contended by some, where the breeder must look for defeat if he expected heredity to come to his assistance in producing fowls of higher fecundity. Where was the luckless breeder to look? Was he to rest on his oars and confess himself beaten?

Before poultry-keeping may become a more profitable and certain business, the egg-laying efficiency of the hen must be increased. The average production of the flock is lower than it should be. How to increase production is probably the greatest of the problems that concern poultry breeders. If they are to secure the fullest measure of success, they must set themselves resolutely to the task of solving this problem. Recent poultry breeding history offers assurance that steady, persistent work will bring rich rewards.

While the interest in poultry breeding has centered largely around the egg, it is only a question of time when the problem of developing meat qualities will command attention. There is an inviting field here for the breeder, one that has barely been touched upon in this country. Notwithstanding the great consumption of poultry, there is only one reason why this is not double what it now is: namely, the poultry that goes to market has not been bred for market; or rather not much consideration has been given to market qualities in breeding. An inspection of the poultry that goes to the average city market amply demonstrates the fact that more attention has been given to breeding for weight than to amount of edible meat on the carcass. There is too large a proportion of bone to meat. What the breeder should aim at, is to increase the amount of edible meat on the carcass without increasing its weight. To accomplish such result would be a worthy achievement for American breeders. The growing demand for greater perfection in meat qualities in the fowl must be met, and one of the developments of the poultry industry that is bound to come in a few years will be a keen competition among poultry breeders to meet the ideals for a perfect table fowl.

PRACTICAL PROBLEMS IN BREEDING

Poultry breeding will never be a business of mathematical certainties. The final result of the breeding must rest largely on the skill of the breeder himself. In other words, poultry breeding is more of an art than a science. The successful breeder, however, follows, consciously or unconsciously, certain laws or principles that have been established or proved by science. A brief explanation of some of these laws follows:

Heredity.—The transmission of qualities or characteristics from parent to offspring is controlled by the law of heredity. Brahma chickens may be hatched from eggs of the same size as Leghorn chickens; the chicks may be the same size when hatched, but from that time on the influence of heredity will be shown in the larger growth of the chick that has an ancestry showing large size. When a pure-bred Leghorn is mated to a pure-bred Leghorn it is almost a certainty that the offspring will be fowls of small size.

When the male offspring begins to crow, it does so because of this same law; its male ancestors for thousands of years have crowed. Sometimes the breeder, and often the nearby neighbor in the early hours of the morning would prefer that this law was more flexible and that it were possible to breed chickens without a crow, but the breeder knows by experience that there are certain characteristics that have become fixed and that if he attempted to change them he would get nothing for his pains. So is egg-laying a fixed characteristic. It is a law of nature or heredity that the hen lays eggs. The law is that like begets like. The practical breeder is guided by this law first and foremost.

But while the law of heredity is persistent and inflexible, while like begets like, there is the strange contradiction in nature that no two individuals are alike. The male chickens all crow; they are alike in that respect, but there are differences in the crow which are easily discernible. So the females are alike in regard to laying eggs; they all lay eggs, but there are differences in the layers; some will lay five eggs a week, others one; some 200 a year, others 20. Another law is seen here.

Variation.—It is the law of variation. This law of variation has already been referred to. Some fowls of the same ancestry or breed vary in number of points in the comb, in size of comb, in length of wattles, in color of eye, in length of limb, in color of plumage, in amount of meat, in size or weight of bone, in number of eggs laid, in size and color of egg, etc. This is variation. Variation is the opportunity for the breeder. The problem that confronts him at the out-

64   POULTRY BREEDING AND MANAGEMENT

set is first to recognize the limitations due to heredity, and, second, to discover wherein certain points or characteristics may be improved by taking advantage of variation.

How may desirable variations be fixed? Is it an evolutionary process? In other words, is it a process of breeding that requires years to accomplish; or may it happen at once?

VARIATION THE OPPORTUNITY OF THE BREEDER
Hen (left) laid 239 eggs in a year.
Hen (right) laid 7 eggs in a year. (Oregon Station.)

If a 200-egg hen be bred from a strain of fowls that lay only 100 eggs a year, will her female offspring lay 200 eggs a year, or will they take after the more remote ancestors and lay only 100 eggs; or will there be a tendency to lay more than 100 due to the influence of the immediate parent? Will the immediate parent transmit her qualities to the offspring, or will the influence of all the ancestors be apparent? Is it a variation that is called continuous because it has been gradually evolved, step by step, or is it discontinuous, appearing suddenly, having none of the characteristics of its immediate ancestors? The old theory of breeding was that all variation was continuous, or, if a sport or mutation did appear, it would suddenly disappear. In other words, all improvement was the result of selection—selecting the best, generation after generation, until finally the desired type or characteristics became firmly fixed. This was the theory of Darwin; but it has been shown that all improvement is not a slow evolutionary process; that it is not all a matter of selection, but that a new type may suddenly appear and start a new variety or a new breed. The vast majority of "sports" or mutants may not breed true; they may disappear as suddenly as they came, but the breeder with a knowledge of the history of new varieties of plants and animals will carefully test any variations that point to higher excellence.

During the past half century at least one breed of fowls owes its existence to the appearance of the mutant. A white "sport" came from a dark breed and resulted in one of our popular breeds. The mutant may disgrace the yard of the fancier who is breeding for uniformity, but the breeder who wishes to perpetuate new and desirable characteristics or establish a new breed must be on the lookout for and carefully preserve such characteristics when they appear.

Two-hundred-egg hens may or may not breed true. Some of them may and some may not, but the progressive breeder will take his chances. So far as it is now known, it is a chance, and the sure way to determine whether one hen or one male will breed true is to test them in the breeding pen.

Reversion.—Variation, however, is not always in the line of progress. Sometimes the offspring may vary away from the line of improvement. Sometimes characteristics that have a counterpart only in remote ancestors, appear suddenly. This is called reversion or breeding back. The scientific name is atavism. What causes reversion is one of the great mysteries. There is a latent tendency, largely unknown, for breeding to evolve backward. It sometimes happens that the offspring shows characteristics that have not been known to appear for centuries in the history of the breed. Reversion works for desirable as well as undesirable traits.

The causes which produce reversion are not always apparent, but factors such as a change in food or climate are known to cause reversion. Characteristics that have apparently been lost, but are not lost, only latent, may reappear when the system of breeding is changed.

The reader is referred to page six for reference to Darwin's experiments on reversion, wherein Darwin claimed that he secured a fowl that reverted to the pattern of the jungle fowl, showing characteristics that had been latent for possibly two thousand years. The experiment was repeated by Dr. Davenport of the Carnegie Institution with similar results.

The point has not yet been reached that the origin of any particular breed or variety may in this way be demonstrated with certainty, but it is possible by crossing a Plymouth Rock, for example, with some other breed, and then recrossing the offspring to discover strong circumstantial evidence as to what breeds were used in producing the Plymouth Rock.

At the Oregon Station a White Wyandotte and a Black Minorca were mated. The offspring were white. Mating the white offspring together chicks were secured that were striped in the down on the back. Neither the Minorcas nor the Wyandotte chicks show stripes in the down. This was evidently a reversion to some remote ancestor, possibly to some of the breeds that were used in the making up of the Wyandottes. It is known that the young of all wild fowls of the Gallinaceous species are hatched with stripes in the down of back, and it is possible that by the proper crossing this characteristic, though latent possibly for a thousand years, would appear in the chick. These stripes still appear in the down of newly hatched chicks of the Brown leghorn breed and some others.

A third cross from the same breeds produce fowls that were barred in the feathers, though barring was not a characteristic of any of the different breeds known to have been used in producing either. It might have been that later, if not in the making of the breeds, by accident, or design, a cross was made with a barred breed, and the characteristic of barring, though latent so long as no crossing was resorted to, reappeared when crossing and intercrossing took place. The fact that a hen may lay only a dozen eggs in a year may be accounted for by reversion to the wild ancestor. In crossing the Barred Plymouth Rock and the White Leghorn, some 90 per cent. of the progeny were white. In crossing the white crosses together, a few of the offspring were blue in color.

RESULT OF CROSSING WHITE WYANDOTTE AND BLACK MINORCA

Female of second generation. Note barring.

Reversion is usually an evil, not always. Where improvement has been going steadily on, reversion must always be an evil. Sometimes, however, progress has gone backward in breeding, and in that case reversion may restore the lost character. The practical lesson to the breeder is that he should eliminate everything in the breeding and management that may cause reversion when he is making satisfactory progress.

The Pure-Bred, or Purity in Breeding.—We are accustomed to considering pedigree as synonymous with purity of breeding. The Shorthorn that can trace an unbroken ancestry back to the Duchess family is a pure-bred. The Berkshire that has a clear line of descent from Longfellow— an animal of superior excellence that belonged to a breeder of Missouri about twenty years ago—is a pure-bred Berkshire. If one of its ancestors had been crossed with a Poland China a number of years ago, even though no trace of Poland China could now be detected, and it would win in the show ring, it would not be pure bred. The pedigree was everything. But Mendelism has given a new meaning to pure bred. It has shown us that purity of breeding has a physiological basis. A bird or animal may be pure bred in respect to one character and inpure in respect to another. A bird, for example, may be pure in respect to the character "comb," but not pure in respect to some other character. If a cock and a hen of a single comb breed when mated breed offspring with single combs, they are pure bred so far as comb is concerned.

If, however, some of the offspring are black and others white they are not pure bred in respect to color of plumage. They would be mongrels in color, but pure bred in comb. It is a question of unit characters, not individuals. Again, if a hen laying 150 eggs in a year, mated to a male from a 150-egg hen, produces offspring that lay less than 100 eggs, she would not be pure bred in respect to egg production. The parents are pure-bred when their own characteristics are reproduced in the offspring with reasonable certainty. A hen is a pure-bred egg producer if she transmits her egg-laying qualities to her offspring, even though her offspring come in various colors, sizes and shapes.

A noted exponent of Mendelism says that purity of type "has nothing to do with a prolonged course of selection, natural or artificial." Again he says: "An organism may be pure-bred in respect of a given character though its parents were cross-bred in the same respect. Purity depends on the meeting of the two gametes bearing similar factors, and when two similarly constituted gametes do thus meet in

BREED IMPROVERS

Pedigreed cockerels, all from stock, with records of over 200 eggs in a year. They were sent to breeders in various sections of the United States and to several foreign countries to breed better layers. Bred at the Oregon Station.

fertilization the product of their union is pure. The belief, so long prevalent, that purity of type depends essentially on continued selection is thus shown to have no physiological foundation. Similarly, it is evident that an individual may be pure in respect to one character and cross-bred or impure in respect of others." (Bateson) Again, "An animal may have one thirty-secondth of the blood of some progenitor, and yet be pure in one or more of its traits."

A fowl may be mongrel in one respect and pure in another.

A Plymouth Rock may be "barred to the skin" and transmit that characteristic to the offspring. It is pure in that respect, but she may not be able to transmit to her offspring the yellow leg color; if so, she is not pure in that respect. The Blue Andalusian is considered a pure breed, and yet from the standpoint of purity it is mongrel so far as color is concerned, for when both Andalusian parents are blue, the offspring are blue, black, or splashed white. Experimenters have found that, on the average, half the offspring will be blue, one-fourth black, and one fourth splashed white. Now when the blacks are mated together the offspring are all black; and white with white gives all white. The blacks and whites breed pure, but the blues are not pure, in other words mongrels, so far as color is concerned. Again, if blacks and whites are bred together the offspring are all blue. No amount of selection, line breeding, or inbreeding, will overcome this peculiarity, or trait of the Andalusian; the Blue Andalusian will remain forever a mongrel race so far as color is concerned.

What then is a Pure-bred Fowl?—It may be defined as one that possesses the characteristics of the breed to which it belongs and reproduces those characteristics in its offspring with reasonable certainty. Purity of breeding refers to the blood lines or pedigree of the fowl and to its ability to transmit the breed characteristics to the progeny. The makers of poultry standards have not been able to incorporate in their standards anything signifying egg-laying points, because it has never been demonstrated that there is any particular shape or type of fowl that indicates its laying qualities, nor has it been possible or practicable up to date to include in the standard a requirement as to performance, or a record of eggs laid. On the other hand, the only requirement to admit a horse to the Trotting Register is speed. He must have speed and an ancestry of speed.

Breeders of dairy cattle have a register of merit, and no animal is admitted to that register without a certified performance record.

If some such plan could be worked out for poultry breeders, it would place utility poultry breeding upon a more certain and profitable basis. It is doubtful if this can be accomplished without state aid. The dairy records are authenticated and certified by state agricultural college officials. It would be possible to maintain breeding stations to which a poultry breeder might send a few fowls to be trapnested for a year, or the state could keep on a farm of its own sufficient stock to furnish at nominal cost a limited amount of stock of good laying pedigree to certain poultry-keepers and farmers in different counties of the state.

BARRED PLYMOUTH ROCK MALE
Son of a 218-egg hen. (Oregon Station.)

This would form a nucleus for a strain of good producers in each community, from which could be sold stock and eggs for breeding purposes in that community, using the state farm from which to secure breeding males with which to maintain the egg-laying qualities of their flocks.

A third method would be for a few reliable private breeders in each county to trapnest a flock and keep pedigree records, selling eggs and breeding fowls only from pedigreed stock. It is not clear, however, how satisfactory progress could be made without state aid or state supervision in some way that will relieve the poultry-keeper or farmer of the burden of keeping the necessary individual records and guarantee the reliability of the pedigree. To be of value this work must be continued year after year. To trapnest the flock for one year and pick out a few of the best layers for breeding, would amount to little. The pedigree to be of value must have several generations behind it, and this means not only that trapnesting must be done each year, but also that a record must be kept showing the individual parentage of the chicks hatched.

A GOOD TYPE OF BREEDER

From 200-egg stock. Note vigor and alertness. Has good show points also. (Oregon Station.)

In any event, there is opportunity for private breeders to do a profitable business, if they have the time to devote to it and the necessary knowledge for the keeping of pedigree records and for the proper mating of breeding stock.

In regard to show standards, it has not been made clear how they can be changed so that the poultry judge in making awards may be able to place proper value on productive qualities as shown by the trapnest record or pedigree of performance. But if the poultry show is to keep pace with the development of productive poultry-keeping, something is due to be done that will change poultry standards and give a different meaning to pure-breds other than that they

PRINCIPLES OF POULTRY BREEDING 73

breed pure or true to certain characteristics of color and shape, but are impure from the standpoint of egg production.

Cross Breeding.—In considering poultry breeding from the farm standpoint, it should not be overlooked that a large proportion of poultry products come from fowls that are not pure-bred. A great many farmers practice crossing; others practice grading — possibly a large majority; while not a few follow another system which may be called mongrelizing. This chapter has to do with the first, crossing. Probably most of the farmers of the country recognize the necessity of introducing new blood into the flock and of avoiding inbreeding, but they have not chosen, intentionally or otherwise, to preserve breed characteristics as they are described in the "Standard of Perfection." Comparatively few of them pay attention to exhibition points. It is contended that the farmer makes a mistake in not keeping strictly standard-bred fowls, but he excuses himself on the ground: First, that the initial cost of stocking up with standard-breds is greater than the business would warrant; second, it has not been demonstrated to his satisfaction that standard-breds are better producers than cross-

BARRED PLYMOUTH ROCK MALE

Oregon Station. Dam laid 214 eggs and sire's dam 218. A fine type of Plymouth Rock, but rather large as a breeder of high egg producers.

breds or grades; third, that the great bulk of animal products of all kinds come from grade stock. Those who are making money out of pure-bred, fancy-bred livestock are those who are in the business partly or wholly of producing and selling breeding stock. Fourth, there would not be enough standard-bred chickens in the country to stock up the farms.

Advantages of crossing.—(1) The crossing of two distinct breeds usually results in greater vigor. This is more apparent where pure-breds have suffered from close breeding or inbreeding. Many breeds of poultry have been injured from too close breeding. It is common history that several breeds, once prominent, are now practically extinct as a result of too much inbreeding. A number of years ago the Black Spanish had a reputation on two continents as a splendid egg-producer, but as a result of insensate breeding for a fancy point it is practically unknown to-day. Its most striking characteristic is its long white face, and the fancier set about making it longer, sacrificing every other point, with the result noted. It may not follow that close breeding is necessarily fatal. It may not be impossible in the hands of expert breeders to intensify fancy points, or any other points, by long-continued inbreeding, without annihilating the breed. It means a great sacrifice in the meantime; its numbers are so diminished, its breeding powers so impaired, that practical poultry-keepers cast it aside.

RESULT OF BREEDING FOR A FANCY POINT

The photograph on the left shows the head of the original Spanish male. The one on the right is the modern bird. (From "Poultry for Table and Market.")

Wright says, "There can be no doubt that too close interbreeding has greatly injured the Spanish fowl, and that not only size, but also constitution and prolificacy have been sacrificed to the white face alone."

Other examples could be given where close breeding for a fancy point has removed breeds from the arena of practical poultry-keeping. The Buff Cochin is another that now gets little consideration from practical poultry breeders, largely because it has been sacrificed by close breeding to the fad of profuse leg feathering. The Brahma has been similarly injured. The Plymouth Rock was in danger from the fad of giving the prize to the bird showing the best barring, other points being given slight consideration, but owing to its wide distribution on the farms of the country and to the fact that there were enough breeders to ignore the extreme demands for barring, it has not suffered as some other breeds have.

The present popularity of the Rhode Island Red is largely due to its vigor, which came from its outgrowth origin, and it would have been better, as McGrew intimates, if many other breeds had been bred on the same plan.

."The effects of too close interbreeding on animals, judging from plants," says Darwin, "would be deterioration in general vigor, including fertility, with no necessary loss of excellence of form." That is, there will be a loss in vigor, but this may not be evident in the form or appearance of the fowl.

"The evidence convinces me," he says again, "that it is a great law of nature that all organic beings profit from an occasional cross with individuals not closely related to them in blood." Again, "The crossing of varieties adds to the size, vigor, and fertility of the offspring."

Edward Brown, in "Poultry-Keeping for Farmers and Cottagers," says: "Recrossing very largely remedies this [deterioration in profitable qualities], for it is found that first crosses between suitable breeds give us hardier and more prolific birds than were either of the parents."

The evils of close breeding of animals are pointed out by Shaler, in "Domesticated Animals." He says: "Among the evils which are to be corrected we may also count that which arises from the unguided development of what are called fancy breeds. Thus among our horned cattle the Jersey has been bred to a point where, from the iniquitous inbreeding, which is against what may be called the morality of nature, they are fearfully subjected to tuberculosis."

"It is a generally received opinion," says Tegetmeier, "that cross-bred chickens are the hardiest and most easily reared."

(2) The use of cross-breds enables many people to engage in poultry-keeping who would otherwise be debarred owing to the comparative scarcity of pure-breds and to the high prices that are demanded for them. If the object is to develop the industry as a means of food supply, it would be a mistake to advocate the slaughter of the cross-breds. If cross-breds were to be eliminated at once, it would mean an immediate and serious decrease in poultry products.

(3) Crossing, where it increases vigor, improves the laying. The productive hen has good vitality. Heavy egg production requires a high expenditure of energy, and to maintain this production the fowl must have stamina. While the loss of vigor may not be apparent in the form or outward appearance of the fowl, it will show in lower production. Vigor is not so essential in breeding for type or for show qualities, but it is very essential in breeding for eggs. In experiments conducted by the writer, a hen weighing three pounds produced 29 pounds of eggs in a year, about ten times her body weight. Another weighing less than five pounds produced 42 pounds of eggs in a year. To with-

stand this strain on her reproductive organs, her vitality must not have been impaired by any system of breeding. To demand production of that intensity from a strain of fowls that have been bred and inbred for generations for any special point is to demand the impossible. It is not here claimed that standard-bred fowls or show birds are necessarily poor layers. It is not impossible for the breeder to breed show birds and at the same time maintain the vitality necessary for the high production, but close breeding for either fancy points or utility points will not insure good layers, any more than the same kind of breeding with mongrels will produce good layers.

(4) New breeds and varieties are produced by crossing. Most of our modern breeds are the result of cross breeding. Crossing induces variation, and it is in taking advantage of these variations that new breeds and varieties arise. Some crosses or hybrids possessing desired characteristics breed true, and the type at once becomes fixed. On the other hand, the great majority will not breed true, and years of careful selection will be necessary to fix the type. The Plymouth Rock, Wyandotte, Rhode Island Red, and other breeds are all the result of crossing. The Orpington, of more recent origin, resulted from crossing many breeds, and William Cook, the originator, said that he got "so many more eggs than he did when the breeds were pure that it gave him a new idea."

Jordan and Kellogg, in "Evolution and Animal Life," say, "Often as much progress can be made in a single successful cross or hybridization as in a dozen or even a hundred generations of pure selection."

The Primus berry was produced by Luther Burbank with a cross between the Siberian raspberry and the California Dewberry. Its fruit excells either parent in abundance and size, and ripens before the two parents begin to bloom.

BARRED ROCK AND WHITE LEGHORN, FIRST CROSS
Male. White is dominant color.

The Loganberry, the product of a cross, is greatly superior to either parent in productivity.

Breeders are just beginning to learn a little as to what may be accomplished by crossing, and it is not unreasonable to expect great improvement in the economic qualities of fowls when breeders master the science and art of crossing. There is an inviting field for developing by crossing new strains or varieties of fowls where egg and meat production forms the chief object sought, but it should only be undertaken by those having skill, experience, and patience.

The first cross will give offspring of one or two kinds: either they will resemble in one or more characteristics one of the parents exclusively, or they will show resemblance to both. Certain characteristics blend; others do not. Where the offspring resemble one parent, and

BARRED ROCK AND WHITE LEGHORN, FIRST CROSS
Female. White dominant color.

the males and females of that cross are bred together, some of the second generation will resemble the other parent, and when these are bred together they will breed true, the offspring will all resemble their parents. This was shown at the Oregon Station in breeding Barred Plymouth Rock and White Leghorn fowls. The first generation were all white, or practically so, taking after one parent. When these crosses were bred together there was reversion to the Barred parent, some of them being barred, and in mating these barred crosses together they bred true and produced only barred offspring. Again, in crossing a White Wyandotte with a

BARRED ROCK AND WHITE LEGHORN, FIRST CROSS
Flock showing dominant color white.

Single Comb White Leghorn, the offspring had practically all rose combs. Breeding the crosses together, the offspring reverted to the Leghorn, some of them showing single combs. Breeding this single-comb offspring together, they bred true to the character single comb. Rose comb is a dominant characteristic, single recessive, and recessives breed true, while the dominants do not. This is Mendelism. A knowledge of these facts will often prove useful to the poultry breeder.

Disadvantages of Crossing.—In the foregoing the advantages of crossing have been enumerated. If the discussion were to stop here it might be inferred that the poultryman must necessarily cross his fowls. But there are certain disadvantages, some of which will now be considered. First, before there can be any crossing, there must be breeds to cross. Why, then, should breeds be made and then unmade? It is not necessary that they should be unmade, if the breeds remain in their original purity. The only excuse for crossing is that breeds have been partly unmade, or they have lost some of their original utility. They may have lost vigor, and size, and productiveness, and the excuse for crossing is to restore those lost characteristics.

WHITE WYANDOTTE—BLACK MINORCA MALE, FIRST CROSS
White plumage color and rose comb dominant characteristics.

There is no need of crossing, however, if sane methods have been followed in breeding. But in crossing, if vigor and fertility be restored, other characteristics will be lost, and it will be for the breeder to decide whether the gain is equal to the loss. For example, he is breeding White Leghorns and they have lost in vigor and productiveness; a cross with Brown Leghorns will restore these, but he has lost the perfection in white color, and it will take several years to eradicate this taint. If color is all important to him, or if other points that may have been lost by crossing are more important than the points gained, then he should hesitate to cross, and depend rather upon the introduction into his breeding yards of birds from other strains of the same breed and preferably from other sections of the country, to improve vigor.

The two alternatives are crossing and outcrossing. The theoretical objection to crossing is that it disturbs blood lines, and the influence of ancestry is lost. In other words, while it may "improve the breed it spoils the blood." While crossing often results in improved strains that excel their parents, causing a tendency upward, it is also true that

THE RECESSIVE COLOR BARRING
Barred Rock and White Leghorn Cross. D621 laid 275 eggs; D. 622 laid 272 eggs in a year. (Oregon Station.)

crossing sometimes reverses the engine of evolution and throws backward. This usually happens when it is continued beyond the first generation. Crossing cross-breds with cross-breds will start the engine going backward; in other words, reversion will happen, and the result is likely to be mongrels, or even a type resembling in some characteristics the wild ancestor. Indiscriminate crossing will lead to degeneracy just as surely as will indiscriminate inbreeding. The first cross will give vigor, as much, probably, as a dozen crosses. While the benefits of crossing cannot be ignored, it must be remembered that the mongrel condition of many farm flocks is due to indiscriminate crossing.

Grading.—Probably the kind of poultry breeding followed by the majority of farmers would be better characterized by grading them by crossing. Grading may be defined as the breeding up of common or mongrel stock by the use of pure-bred sires. The object is not to restore lost vigor or other lost characteristics, nor to establish new breeds, but to improve the flock by means of the sire only.

"The failure to make the most of grading," says Davenport in "Principles of Breeding," "is the largest single mistake of American farmers." The great bulk of cattle that furnish the meat supply of the world are grade Shorthorns and Herefords. The same thing is true of farm poultry-keeping, the failure to make the most of grading by the use of pure-bred males is the farmer's greatest single mistake. In four or five generations, by the use of pure-bred males, a variegated mongrel-looking lot of chickens may be bred up to a uniform type resembling closely the breed to which the male belongs. If the male is chosen, however, as he naturally will be, from a strain of heavy layers, the farmer will have the satisfaction and the pleasure not only of receiving greater profit as a result of his labor, but also of witnessing an object lesson in breeding of supreme interest through the gradual but sure realization of his ideals both in increased production and in the gradual unfolding of a distinct type and color.

The important thing in grading is to begin with an ideal and stick to it. If the result sought is higher egg production, the breeder should use preferably one of the smaller breeds, but certainly a strain that can show records of high production. Under no condition should this purpose be departed from. If uniformity of excellence in laying is desired, the object will be quickest secured by using each year a male that has a good pedigree in that respect. Changing the breed or type of male each year will result in getting nowhere. The failure to make the most of grading has been due to the occasional or frequent use of grade males.

A grade may have apparently all the characteristics of the pure-bred; he may look so attractive that the breeder is tempted to take a chance and use him for breeding, with the result that improvement is likely to go backward. The grade may himself have all the characteristics, but he has not the ancestry or blood lines behind him to insure the transmission of those qualities, and instead of grading up, the process is liable to become mongrelizing. It should also be clearly understood that the male should not only be pure-bred, but he should have that purity of breeding that extends to egg-laying; in other words, he should be from a strain that is known to consist of good layers.

Prepotency.—The reason a grade male may not transmit his characteristics has been ascribed to a lack of what is called prepotency. Prepotency, therefore, is the ability of the parent to fix his characteristics in the offspring. All parents have not this power in the same degree. This is the significant fact for the breeder. It is not enough to know that a certain male or female has a long pedigree, or that the blood lines have been carefully preserved for years. Prepotency does not always follow blood lines. Later knowledge has given a modified meaning to prepotency, and the fact has been proved that one individual may be prepotent and another of the same blood lines, possibly of the same parentage, is not. It is true that the individual having been bred pure to a certain type for a great many generations is more likely to transmit his characteristics than would a grade; the chances are much in favor of the pure-bred, otherwise all the laws of heredity would be of no avail. But it is certain that fowls of the same breeding are not equally prepotent. The Mendelian law of dominance furnishes the explanation.

Dominance.—There are certain characters that are dominant. For example, white plumage is a dominant color. This was shown in the cross mentioned between Barred Plymouth Rocks and White Leghorns, where the offspring were all white. They resembled only one parent in color. Color, however, is only one character. The offspring of a cross may take after one parent in one point and the other parent in some other point. For example, in crossing a black Wyandotte with a White Leghorn, the offspring will resemble the Wyandotte in the kind of comb, but the Leghorn in color of plumage, a white chicken with a rose comb. It is said, then, that the white color and the rose comb are dominant characteristics, and that single comb and black color are recessives, and no matter what breeds may be crossed, the dominant characters of white plumage and rose comb will show in the offspring, and the recessive characters will not. It is not a question of ancestry or blood lines.

While the offspring have all rose combs they are impure rose, and this is brought out in the next generation, when the cross-breds are bred together. In this generation reversion takes place, and on the average 25% of the offspring will have the recessive character of single combs; 75% will have rose combs; but of these, 25% are pure rose, that is, they will forever breed pure to rose combs, but the 50% are impure. This 50% when recrossed will segregate in the same way, 25% single combs, 75% rose combs, but of the 75% only 25% will breed pure rose combs, and so on.

It resolves itself into a question of testing the breeding powers rather than a question of selection. The 25%

recessives will breed pure single combs and no amount of selective breeding will make them purer. This is the discovery of Mendel. Twenty-five per cent of the dominants, rosecombs, will breed pure; that is, a third, but the only way to pick out the third or determine which will breed pure rose combs, is to test them by breeding. The way to test them is to mate them to single comb fowls, and if the progeny have rose combs, the rose comb parent is pure and will always breed pure rose combs. This requires, of course, individual mating. All this gives a new meaning to prepotency. Prepotency is not altogether dependent upon length of pedigree.

Egg Color and Dominance.—It has been discovered, however, that not all characters segregate in this way, and herein comes some confusion. Some characters segregate, others unite or blend. In place of resembling one parent, the offspring resembled both in part. Again referring to the experiment of Barred Plymouth Rocks and White Leghorns, the color of eggs laid by the female offspring of the cross showed a blend. The color of the Leghorn egg is white, while the egg of the Plymouth Rock is brown. The eggs from the cross averaged medium, showing the influence of both parents. This character, color of egg, did not act like the character color of plumage. In the one case there was a blend, in the other, segregation. In one, the offspring took after both parents, in the other, after one parent,

Pure-bred not Always Prepotent.—In carrying this experiment further, white cross-bred pullets from the mating of Barred Plymouth Rocks and White Leghorns, were mated to pure-bred Barred Plymouth Rock males, with the result that the offspring had white plumage. They took after the cross-bred parent, rather than the pure-bred parent. Other white pullets of the same cross were mated to pure-bred Brown Leghorn males, and out of over 150 chicks from this mating less than a dozen took after the pure-bred male in color; the balance were all white. The practical point brought out in this experiment is that prepotency does not follow blood lines, and that in mating a cross-bred or mongrel to a pure-bred it does not necessarily follow that the offspring will take after the pure-bred in all characters. Prepotency is not always measured by length of pedigree

While Mendelism is yet in an "embryonic" stage, and while confusion prevails as to its teachings in certain respects, a clue has undoubtedly been found that will lead the way to important developments in the future.

Outcrossing.—The term outcrossing is frequently used by breeders. It means the use of males from strains different from those of the females, but belonging to the same breed or variety. It is breeding within breed lines, but not within family lines. If a careful system of outcrossing be practiced, resort to crossing will probably seldom be necessary to keep up the stamina of the breed. This is the main purpose of outcrossing. Another object in outcrossing is to improve the family or strain in some point or character which it may lack by introducing blood from another strain which is strong in the character the other lacks.

Inbreeding.—The mating together or breeding together of closely related males and females, is inbreeding. Where the relationship is close, inbreeding is the term used, but where the relationship is more or less remote it is called by many authorities line breeding. The only difference, if there should in reality be any difference, between inbreeding and line breeding, is a difference in degree of relationship. It is doubtful whether it would not be just as well to call it all by the one name, "inbreeding."

Breeding of brother and sister together is the closest kind of inbreeding. Mating parent to offspring is also close breeding, though this is frequently called line breeding. The subject of inbreeding is a much debated one, and until there is a better and more perfect understanding of its effects the debate should continue. It is unquestionably a most important problem.

There are those who steadily maintain the ground that inbreeding is necessary to breed improvement. It is probably true that most of our breeds of poultry, as well as of live stock, were largely inbred in the making. It may be open to doubt whether this was not largely due to circumstances or to the fact that in the making of new breeds there were not at hand two or more families unrelated by blood lines from which to draw upon, rather than to the merit of the system itself.

It is undoubtedly true that close breeding or inbreeding has been a costly blunder, and it is playing with a dangerous weapon when inbreeding is held up to poultry breeders as always desirable or necessary. It is a problem that can be solved only after long experiment, and it appears to the writer that the data is not at hand upon which to base final judgment. In the meantime a common-sense view should be taken by the breeder, which should prompt him to avoid close breeding and suggest that where it seems necessary to fix or maintain some desirable characteristics, the breeding together of distant relatives may possibly be practiced to advantage.

The purpose of the breeding should be considered in discussing the effect of inbreeding. The evil effect of inbreeding may not be apparent in the form or beauty of the fowl, but may result in reduced vigor and lower breeding power. It may show in the egg yield, in the fertility of the eggs, and in the vigor or mortality of the chicks, but not necessarily in the type or prize-winning qualities. Again, it may be possible to breed a larger proportion of prize winners through inbreeding than by outcrossing, but at the same time the egg-laying characteristics may be injured. The explanation is that the reproductive organs, i.e., the egg-laying organs, are more closely related to vigor than is the shape or type of fowl.

It is denied by many that a loss in vigor necessarily results from inbreeding. By careful selection of breeding stock, it is claimed, no loss in vigor will follow. Let us see.

What is the Purpose of Inbreeding.—It is to fix desired characteristics. If a superior fowl be found, one that possesses in a high degree certain points of value to the breeder, it is claimed that by breeding her to her son her characteristics will be more quickly fixed in the offspring than would be the case by any other system of breeding. The points of superiority may be color of plumage, shape of comb, shape of body, or number of eggs laid, and various other points that the breeder wishes to fix. If the hen has proved to be a good layer, the theory of inbreeding is that by breeding her to her son, there will be more probability of getting good layers than if she be mated to a male that is not related to her.

If the point bred for be color or type, it may be that inbreeding or line breeding will give a larger proportion of offspring strong in those points than would outcrossing, even though the males in either case be equally good in those points. But it is a different matter when the point to be bred for is one that has to do with their productive or reproductive qualities, because those points are so intimately related or correlated to vigor of the fowl that it is doubtful if the theory will hold.

Inbreeding Experiments.—Recent experiments at the Oregon Station indicate that the evil effects of inbreeding overbalance the possible good. Fowls with no apparent lack of vigor, and no defects in external points of shape and color, showed, first, in a lower fertility of eggs, in a lower percentage of fertile eggs hatched, and in a higher rate of mortality in the chickens, that there was a loss of vigor due to inbreeding mother to son. Second, the result showed

OREGON STATION HEN C543, AND THE EGGS SHE LAID IN A YEAR—291

An exceptional layer though inbred.

decreased egg yield in the pullet offspring. From different matings the inbred pullet offspring showed a lower average egg yield than other matings not inbred.

Again, it is claimed that inbreeding tends to uniformity

of type and that it "discourages variability." ("Principles of Breeding," Davenport, p. 610.) That may be true of type, but these experiments with fowls gave results in production exactly the opposite. There was less uniformity in egg production from those inbred than from those not inbred. An inbred pullet, a daughter of hen 250 (record 402 eggs in two years) inbred to her son, laid 291 eggs in a year, one of the most remarkable layers ever produced at the Station. The same mating that produced this phenomenal layer (291 eggs) produced the second poorest layer in the flock, which laid 124. The average of all the inbred pullets of this mating was 181 eggs.

The same thing was shown in the production of pullets of another mating, hen 034 (record 229) inbred mother to son—greater variability and lower production than the other matings. The results in this case showed a high record of 237 eggs, and a low record of 119, the lowest of all the flock, the average being 187 eggs. Those of the flock inbred averaged 182, against an average of 219 of all not inbred, or 20% more. In the latter case, the highest was 303 and the lowest 163. In the previous year, the daughters of 034, not inbred, averaged 210 eggs, and the daughters of 250, not inbred, averaged 221. In all these matings males from high producing hens were used. The egg yield from the inbred fowls while 20% lower than the others is above the yield of the average flock of fowls, indicating that a good yield may be secured by inbreeding, not because of it, but in spite of it. It will pay to inbreed some, rather than use breeding stock that are indifferent or poor layers. So far as fixing the character of egg production, inbreeding proved a failure. Not only was there lower production but there was greater variability in production.

If as these experiments strongly indicate that heavy egg production demands a high vigor and that the reproductive organs of the fowl are peculiarly sensitive to inbreeding, then the breeder whose object is higher egg production must not follow inbreeding, even though it may have accomplished in the way of improving type in both live-stock and poultry all that its strongest supporters claim for it.

CHAPTER VII

PROBLEM OF HIGHER FECUNDITY

Various factors relating to environment, such as feeding, housing, and management, affect the egg yield, and are discussed in other chapters. Unless these conditions are favorable the egg yield will be low. It has also been seen that systems of breeding and mating that affect vigor influence the egg yield.

It is another question whether high egg-laying is trans-

LIKE BEGETS LIKE

Two full sisters. Oregon Station. C119 (left) laid 241 eggs. C166 (right) laid 233 eggs.

mitted from parent to offspring. This is probably the most important problem of all. It would seem that any doubt on this point could very easily be set at rest by actual demonstration, but experimenters have found the problem a difficult one. The actual experimental data at hand are not very extensive and possibly not conclusive enough to

LIKE DOES NOT ALWAYS BEGET LIKE
Two full sisters. C48 (left) laid 268 eggs. C60 (right) laid 3 eggs.

satisfy scientific demand in regard to the mode of inheritance, if not of the fact of inheritance itself.

The breeding work of the Maine Station, begun by Professor Gowell and continuing nine years in co-operation with the Bureau of Animal Industry during part of that period, produced negative results so far as raising the standard of egg production of the

OREGON STATION HEN D18
271 eggs in a year. From same dam as C119 and C166, but different sire. Note short body.

flock under investigation was concerned. After a thorough study had been made of the trapnest records the announcement was made in 1908 (Bulletin 157) that "there is no evidence of any increase in the average production of the flock." The average production and the number of fowls in the experiment are given herewith for the different years:

Year and Pen		Number of fowls completing the year	Actual average production per hen
1899–1900		70	136.36
1900–1901		85	143.44
1901–1902		48	155.58
1902–1903		147	135.42
1903–1904		254	117.90
1904–1905	50 bird pens	283	134.07
1905–1906	50 bird pens	178	140.14
1906–1907	50 bird pens	187	113.24

In a subsequent bulletin (Number 166, 1909) it was stated: "The aim so far has been to set forth in as clear and unequivocal manner as possible the definite fact that in the Station's experience thus far the daughter of a 200-egg hen is on the average an exceptionally poor winter layer instead of an exceptionally good one." In this case the mothers' average winter production, November 1 to March 1, was 55.8 eggs, and the daughters' 15.29 eggs per fowl. The results further showed that the daughters of hens laying less than 200

BARRED ROCK HEN A78
Record 212 eggs. A good type of Plymouth Rock.

eggs gave a higher winter egg production than those from the 200-egg hens.

Again, in Bulletin 192 (1911), the following statement is made: "There does not exist any critical evidence that the selection of the highest laying birds on the basis of the trapnest record as breeders will insure or guarantee any definite permanent improvement in the average flock production."

"It now seems quite generally agreed," quoting from the same bulletin, "that about the only profitable function of the trapnest in practical or commercial poultry-keeping is in connection with special needs or problems, as, for example, in the work of the fancier."

A GOOD PLYMOUTH ROCK HEAD, WITH THE STAMP OF VIGOR
Son of A78—212 eggs. Sire's dam 259 eggs. Oregon Station.

The publication of these results was somewhat discouraging to breeders and provoked widespread discussion. The critical reader will observe, however, that the failure to show an improvement in production by selection is not put forward as proof that it is impossible, but only that there is no evidence in the records of that Station that it is. Again, in the case where the daughters of 200-egg hens were poorer layers than the daughters of layers not so good, it is not held that this result must always be expected, but only that " in the Station's experience" this was obtained.

The statement in regard to the value of the trapnest is rather difficult to understand because all improvement must rest upon individual performance. A knowledge of individual performance is only possible where the trapnest is used.

It seems to the writer that the extreme difference in the yield of 55 eggs as the winter production of the dams and 15 eggs for the daughters, cannot be satisfactorily explained except upon the theory that environmental conditions were unfavorable in the case of the daughters — that the conditions surrounding their breeding and management were in some way unfavorable to high production.

BARRED PLYMOUTH ROCK HEN 65— 218 EGGS

Oregon Station's first 200-egg hen. Her daughter and granddaughter on p. 97.

The methods followed in selecting the breeding stock, in the nine years' experiment, was to use only hens that had records of 150 eggs or more, and after the first year male birds only were used whose dams laid 200 eggs or more.

At the Oregon Station, later experiments produced different results. The records of six years' breeding work with Barred Plymouth Rocks are summarized on the next page.

The original flock of 95 pullets were purchased from six different breeders. The first and second years' results have little significance so far as the question of inheritance is concerned. There had not been time enough to make selec-

PRODUCTION IN FIRST YEAR OF LAYING

Year	Number of hens	Laid first year	Highest	Lowest	Per cent laying 200 eggs or more
1908–09	95	84.7	218	6	1.05
1909–10	28	121.2	183	70	0.00
1910–11	43	164.6	259	6	25.58
1912–13	108	179.2	268	3	22.22
1913–14	160	176.5	271	7	23.12

tions of breeding stock on the basis of individual production. In the third year the pullets were from the original stock, except that a third of the poorest layers were discarded at the end of the first year. This flock of pullets, however, had not individual pedigrees. The fourth year records are not given as they were not fairly comparable with those of other years. In the fifth year the pullets, 108 in number were all from pedigreed high producers.

A116—DAUGHTER OF 65
Laid 218 eggs.

D172, GRANDDAUGHTER OF 65
Laid 221 eggs.

The average production of the dams was 202 eggs in a year. Their sires were from dams, one with record of 218 eggs, the other 169. In the sixth year there were 160 pullets which averaged 176.5. Their dams averaged 187.9, sires' dams 219.8 and dams' dams 211.7. Hens that died are eliminated in the calculation for each year.

No "new blood" was introduced during the six years, but inbreeding was avoided. The parent stock was selected

A122—MOTHER

Laid 259 eggs in a year. Some of her daughters and granddaughters are shown on following pages. All of medium size and, with one exception, are short in body.

each year on the basis of trapnest records, no attention being paid to shape or type. The breeding fowls, however, represented fairly well the general characteristics of the breed.

Were further evidence needed as to the inheritance of fecundity it is brought out in the table on p. 102 compiled from the Oregon experiments, in which it will be seen that the progeny of selected high layers produced 207.3 eggs per hen, while the progeny of selected poor layers averaged 138.1 eggs. Male X of unknown ancestry, mated to poor

PROBLEM OF HIGHER FECUNDITY

C146—209 Eggs

D84—236 Eggs

D90—240 Eggs

D177—268 Eggs

A FAMILY OF HIGH PRODUCERS—DAUGHTERS AND
GRANDDAUGHTERS OF A122

D118—233 Eggs   D119—209 Eggs

GRANDDAUGHTERS OF A122

A77—214 Eggs

A GOOD TYPE OF PLYMOUTH ROCK AND A PRODUCER OF GOOD LAYERS

Three daughters on opposite page, and two granddaughters on page 102.

PROBLEM OF HIGHER FECUNDITY

layers, produced daughters that averaged 117.1 eggs each. Mated to good layers the daughters averaged 179.7 eggs. The same result is shown in the case of the sire whose dam laid 218 eggs.

Selection and Cross Breeding. — Is the problem of higher fecundity a question of selection altogether, or is it a question of constitutional vigor alone? Does the work of the breeder begin and end with selection, or does it begin and end with vigor in the stock? It is well known that many strains of pure breeds, due to close breeding or other causes, lack in vigor. It is also known that crossing two breeds or varieties will restore the vigor lost by close breeding, and

D39—270 Eggs

C58—205 Eggs

D74—244 Eggs

DAUGHTERS OF A77

D52—214 Eggs   D106—225 Eggs

GRANDDAUGHTERS OF A77 (on page 100)

EGG PRODUCTION OF PULLETS FROM BOTH GOOD AND POOR LAYING ANCESTORS

Year	Dams laid first year	Number	Sire's dam laid	Daughters	Number
1912–13	20	1	169	144.4	5
1912–13	74	1	218	136.5	2
1913–14	91.9	7	20	156.1	9
1911–12	122	2	*	117.1	9
Average				138.1	
1911–12	215.5	2	*	179.7	28
1912–13	187.7	9	218	185.2	67
1912–13	226		229	215.6	32
1912–13	402‡	1	402‡	187	3
1913–14	385‡	11	204	223.7	50
1913–14	211.8	4	259	185.5	28
Average				207.3	

*Male is of unknown pedigree.
‡Two years' record.

this is accomplished in one cross. Experiments at the Oregon Station showed that crossing Barred Plymouth Rocks and White Leghorns gave a decided increase in yield, but a still more decided increase was obtained by following up the crossing with selection. Crossing alone was not sufficient. Only by high constitutional vigor, aided by selection, can the highest production be secured.

WHITE LEGHORNS FIRST YEAR, CROSSES IN SUBSEQUENT YEARS
PRODUCTION FIRST YEAR OF LAYING

Year	No. of hens	Laid first year	Highest	Lowest	Per cent laying 200 eggs or more
1908–09	50	106.9	183	2	0.00
1909–10	63	135.6	211	14	4.76
1910–11	39	149.5	257	28	15.38
1912–13	23	218.2	303	124	69.52
1913–14	50	223.7	278	92	70.60

In these experiments 50 pure-bred White Leghorns averaging 106.9 eggs were crossed promiscuously with Barred Plymouth Rocks that averaged 84.6 eggs. The cross-breds in the first generation averaged 135.6 eggs. Breeding them back to pure-bred Leghorns the pullets with three-fourths Leghorn blood, averaged 149.5. In the latter flock of crossbred hens, those with records averaging 208 eggs were selected for breeding. Five of them were mated to males from a Leghorn hen with record of 229 in her first year and 407 in two years. Three were mated to a cross-bred male of three-fourths Leghorn blood whose dam laid 402 eggs in two years. The daughters from these matings averaged 218.2 eggs, one of them laying 303 eggs. In the next year (breeding females of the same grade were used— three-fourths Leghorn blood, average production 385 eggs per hen in two years, dams' dam 402 eggs in two years. These were mated to a pure-bred Barred Rock male whose dam laid 204 eggs in one year and sire's dam 218. The

C35—220 Eggs
Daughter of A79

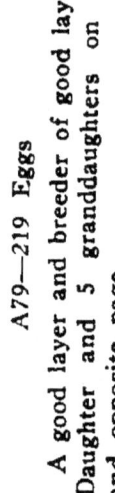

A79—219 Eggs
A good layer and breeder of good layers. Daughter and 5 granddaughters on this and opposite page.

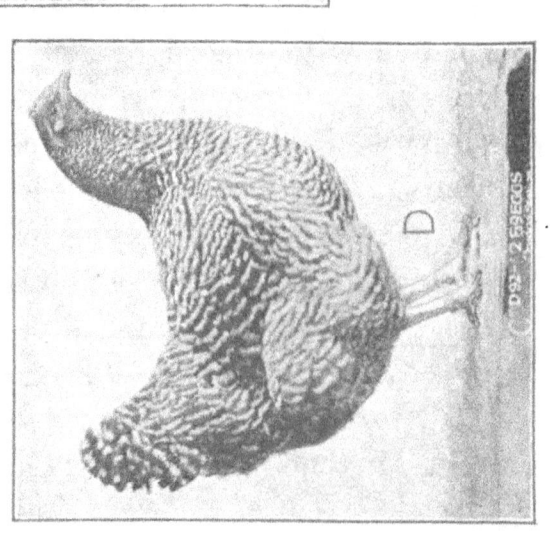

D92—259 Eggs
Granddaughter of A79

PROBLEM OF HIGHER FECUNDITY   105

resulting flock of 50 pullets—five-eighths Barred Rock and three-eighths Leghorn blood—averaged 223.7 eggs.

These and other results secured by the writer at the Oregon Station indicate clearly not only that high egg-laying is transmitted, but that vigor and selection are both

D166—217 Eggs   D34—234 Eggs

D180—239 Eggs   D5—233 Eggs

GRANDDAUGHTERS OF A79

necessary if the highest results are to be secured in production.

It should not be assumed from this experiment that crossing of different breeds is always necessary to secure vigor. The same result may be secured by crossing different strains of the same breed, in other words by outcrossing.

Mode of inheritance.—Granting that the fact of transmission of high egg-laying has been demonstrated, there remains the further question as to the mode of inheritance. Does it come about all at once according to the Mendelian law of dominance and recessiveness, or is it an achievement that comes bit by bit after years of patient selective breeding? Is high egg production a sex-limited affair in its inheritance? Is it inherited from the dam and dam alone, or does it come through the sire and sire alone?

So far as the Oregon investigations have gone, the results do not bear a Mendelian interpretation. They do not show that high egg production is either dominant or recessive to low production. When high producers were mated to sons of high producers the daughters were neither all high nor all low producers. Mating high producers together, the daughters did not equal the production of the parents on the average. When low producers were mated the daughters did not take after either or both of the parents, but showed a higher egg production than the dams or sires' dams. In the one case there is a pulling down, in the other a pulling up to a general level. Apparently the daughters do not take the characteristics of the mother to the exclusion of the sire's mother, or the reverse.

It appears as though high egg production is the accumulated result of the selection of high production breeding stock carried on for many generations. The breeder, however, will make rapid progress in reaching the high standard in proportion as he is successful in identifying the exceptional individuals that possess in a high degree the power of transmitting desired characteristics to their offspring.

Is High Fecundity Sex-Limited.—In other words, is high fecundity inherited through the sire alone, or from the dam alone? This question has been the subject of investigation by Doctor Pearl of the Maine Station, and his conclusions are given in these words (Maine Bulletin 205): "High fecundity may be inherited by daughters from their sire, independent of the dam." . . . "High fecundity is not inherited by daughters from their dams."

WHITE LEGHORN HEN O34

229 eggs in first year, 786 eggs in five years. Two daughters below. Oregon Station.

B12 251 eggs in 1 year

C551 607 eggs in 3 years

DAUGHTERS OF O34

OREGONA—OREGON STATION WHITE LEGHORN HEN A27

987 eggs in five years. At beginning of sixth year had laid more than 1,000 eggs. The greatest long-distance trapnest record known: first year, 240 eggs; second year, 222 eggs; third year, 202 eggs; fourth year, 155 eggs; fifth year, 168 eggs.

. . . "A low degree of fecundity may be inherited from either sire or dam." To state Doctor Pearl's conclusions in another way: High fecundity may be inherited from the sire, or may not. In other words, some sires will transmit this characteristic and some will not. Whether nine in ten, or one in a hundred have this power, we are not informed, and Doctor Pearl does not presume to know. Breeders therefore will not be misled into the belief that all males have the power of transmitting the egg-laying characteristics of their dams. Pearl, however, is definite when he says that "high fecundity is not inherited by daughters from their dams."

The Oregon experiments which have shown a remarkable increase in production, with strong evidence that it was due to selective breeding, do not appear to show that it came in a sex-limited way. They show, on the average, that both sire and dam exert an influence, but that the influence is not confined to the immediate parents. It is true that some males, as well as females, have a greater prepotency or power of transmitting fecundity than others, but it cannot be said, so far as these experiments have gone, that it comes only from "one side of the house." Table on p. 102 may be studied in this connection.

Progression and Regression.—The production of the progeny never reaches that of the parent stock when the egg production of the parents exceeds the average of their generation. There is, however, a progressive increase each year when the parents have been selected among the individuals that have production records higher than the average of the flock. (See table p. 111 Oregon Station experiments.) This is the principle or law of progression.

There is another principle or law operating in the other direction; that is regression. There will be regression or a decrease in production unless the breeding stock be selected among the highest producers. When no selection of any kind is practiced, the tendency is downward. The average of all the ancestry is pulling backward. Selection is necessary if the breeder is to do no more than maintain the standard of production. He cannot "rest upon his laurels" without going backward. He must select and continue to select.

Variability *versus* Uniformity.—High excellence is not correlated with uniformity of production, as the Oregon experiments show (page 111). Breed improvement does not mean that all individuals of the flock are bred up to the same level of production. It is not a leveling process. The gulf between the high and the low individuals is not bridged by selective breeding. The experiments indicate that breed improvement, so far as egg production is concerned, means the raising of the standard of production of the individual. In other words, variability does not decrease with improvement in production. There are fewer poor layers as a result of selection and more good ones, but the range between the high and the low remains practically the same. In the case of the Barred Plymouth Rocks the mean of production moved up from between 61 and 80 in the first year to between 161 and 180 in the sixth year. In the case of the Leghorns and crosses, practically the same law is shown.

Breed improvement, therefore, depends upon raising the mean or average production at the same time as the maximum production is raised. This is what happened in the Oregon experiments. The maximum individual production was raised each year while the average of the flock was also raised. As the average production of the flock is raised, the probabilities are that individual high records will increase in like manner. The true breeder, therefore, will ignore a fixed standard of production and breed for a progressive increase, and no one can yet say what the maximum production of the hen is. (See page 111.)

Hen's Potential Capacity.—That the conditions under which a hen lives affect her egg yield and determine, in a measure at least, her degree of fecundity, is a truth discussed elsewhere. This is supported by investigations made by Pearl as to the number of oöcytes (eggs) in the hen's ovary. It is apparently not from lack of eggs or oöcytes in the hen that the egg yield is low, for the count

BARRED PLYMOUTH ROCK HENS GROUPED ACCORDING TO PRODUCTION, SHOWING PROGRESSIVE INCREASE

Number of eggs laid	1908–09	1909–10	1910–11	1912–13	1913–14
1–20	1	..	2	1	1
21–40	6	..	..	..	..
41–60	19	..	1	..	..
61–80	27	5	1	..	2
81–100	17	3	..	1	6
101–120	11	8	..	4	11
121–140	7	6	8	10	15
141–160	4	2	6	15	22
161–180	1	3	9	27	41
181–200	1	1	6	27	26
201–220	1	..	9	14	18
221–240	..	..	1	7	11
241–260	..	..	..	1	4
261–280	..	..	..	1	3
281–303	..	..	..	..	..
Total hens	95	28	43	108	160

HENS GROUPED ACCORDING TO PRODUCTION SHOWING PROGRESSIVE INCREASE

(First year, White Leghorn hens; subsequent years, Crosses.)

Number of eggs laid	1908–09	1909–10	1910–11	1912–13	1913–14
1–20	3	2	..	..	..
21–40	2	..	1	..	..
41–60	1	..	..	..	..
61–80	6	..	2	..	..
81–100	8	6	3	..	1
101–120	9	15	4	..	..
121–140	9	16	5	1	..
141–160	5	9	6	1	1
161–180	5	8	12	1	4
181–200	1	5	3	5	6
201–220	..	2	2	7	16
221–240	..	..	..	2	6
241–260	..	..	1	2	9
261–280	..	..	..	2	7
281–303	..	..	..	2	..
Total hens	50	63	39	23	50

showed in some cases, even in poor laying hens, the presence of over 2,000 oöcytes.

So far as the number of eggs in the ovary is concerned, hens are all "born" with the inherited tendency to lay. The lowest number in any one hen, as reported in Maine Bulletin 205, was 914; the greatest number 3,605. By using a low-power dissecting lens to aid the eye, the enormous number of 13,476 oöcytes were counted in one hen's ovary.

It should be understood that the ovary of the hen, even before she lays any eggs, contains all the eggs, called oöcytes, that she will ever lay. More than that, she has many times more eggs than she will ever lay. Why doesn't she lay them? That is the problem. Is it a lack of inherited ability to lay, or is it because of improper feeding and care? Is it the business of the poultryman to so mate his fowls that the ability to lay the greatest possible number of eggs will, in some manner, be transmitted from parent to offspring? Or is it his business to so feed and house the hen, in other words put her under such favorable environment, that she will empty her egg reservoir, so to speak, during her natural laying life

The poultryman who is gifted, however, with the faculty of using common sense, will not neglect either the breeding, the feeding or the care and expect to get the largest possible egg yield. A knowledge of the fact that the hen has a potential possibility of several thousand eggs, strongly emphasizes the importance of environmental factors, in other words, good feeding, proper housing and care.

Actual Limit of Production.—Before the count of the oöcytes had been made the idea was somewhat prevalent that 600 eggs was the limit of production of a hen. This theory seems to have originated with a French writer named Geyelin, who said: "It has been ascertained that the ovarium of a fowl is composed of 600 ovula or eggs; therefore a hen during the whole of her life cannot possibly lay more eggs than 600, which in a natural course are distributed over nine years in the following proportion."

This has been abundantly disproved by trapnest records. At the Utah station, prior to 1905, a number of egg records

B42—CROSS-BRED HEN, LEGHORN BLOOD PREDOMINATING

Laid 834 eggs in four years at Oregon Station. A world's record for four years. First year, 228 eggs; second year, 250 eggs; third year, 184 eggs; fourth year, 172 eggs.

A60—CROSS-BRED HEN, ¾ LEGHORN AND ¼ PLYMOUTH ROCK

816 eggs in four years and 958 eggs in five years.

were secured exceeding the 600-egg limit in less than four years of laying. (Bulletin 92, by the writer.) Since then one hen has laid 816 eggs in five years. At the Oregon station the writer has secured many records exceeding 600 eggs. In one case 664 eggs were laid in three years, and 819 in four years by the same hen. In her fifth year she reached a total of 987 eggs. At the beginning of her sixth year she passed the 1,000-egg mark. Another laid 958 in five years. Remarkable are the records of hens, B42 which has laid 834 in four years, and of B14, 827 in four years. If they continue in their present condition of health, all these hens should next year reach and pass the 1,000-egg mark. These are the highest authenticated four- and five-year records known.

Long Distance Laying.—Some idea of the possible limit of production is given in the following table, which records the egg yield of 12 sisters at the Oregon Station for three years:

"LONG DISTANCE" EGG RECORDS

Hen No.	First year	Second year	Third year	Totaled
B4	217	214	172	603
B8	246	160	159	565
B13	206	226	206	638
B14	215	206	208	629
B170	226	220	177	623
B177	193	212		
B213	198	224	230	652
B222	188	199	231	618
C425	235	199		
C543	291	150		
H81L	161	194	138	493
H53N	168	196	173	537
Average	211.1	200	188.23	593.3

Dam's record: First year, 200 eggs; second year, 202.

These hens laid an average of 211.1 eggs in their first year, 200 in their second year, and the remaining nine hens in their third year averaged 188.23. The total average production for the three years was 595.3 eggs each. The dam of these pullets laid as many eggs the second year as the first, and it would appear that this characteristic in the dam was transmitted to the daughters in a noticeable degree.

The high limit of production is further shown in the following compilation from the Oregon experiments.

OTHER "LONG-DISTANCE" EGG RECORDS

Hen No.	First year	Second year	Third year	Fourth year	Fifth year	Sixth year	Total	Dam's record
B12	251	152	..	..	..	..	403	229
C551	214	203	190	..	..	..	607	229
H81N	142	224	179	..	..	..	545	229
B2	223	203	188	..	..	..	614	155
C457	250	175	..	..	..	..	425	..
C459	261	191	..	..	..	..	452	..
C463	232	202	..	..	..	..	434	..
B42	228	250	184	172	..	..	834	..
034	229	178	140	144	95	6	792	..
A27	240	222	202	155	168	..	987	..
A60	177	234	226	179	..	..	816	..
C504	243	178	..	..	..	..	421	240
C515	241	182	..	..	..	..	423	240
C589	211	204	..	..	..	..	415	240
C516	267	174	..	..	..	..	441	215
C519	272	182	..	..	..	..	454	205
C552	217	185	..	..	..	..	402	209
C483	265	145	..	..	..	..	410	177
C512	217	217	..	..	..	..	434	177
C490	231	216	..	..	..	..	447	257
C521	303	209	168	..	..	..	679	201
C547	250	225	214	..	..	..	689	201
C90	215	188	..	..	..	..	403	191
Average	233.86	197.34	188	159.33				212.2

Laying Longevity.—The more important point brought out in these long distance records, however, is the evidence that the period of longevity, or the profitable laying period, may be considerably lengthened. The third year record of the 12 sisters—188.2 eggs—is remarkably good laying for first year hens or pullets, and considerably higher than the average flock of pullets. The average or unimproved flock of hens does not pay for its keep after the second year when eggs are sold for market purposes. If by proper breeding this period could be lengthened to four years, it would mean that once in four years, instead of once in two years, the flock would need to be renewed, thus cutting out half

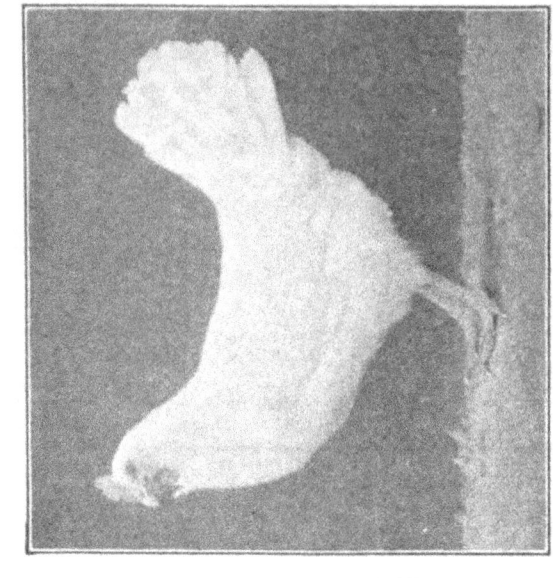

QUEEN UTANA
816 eggs in five years. (Utah Station.)

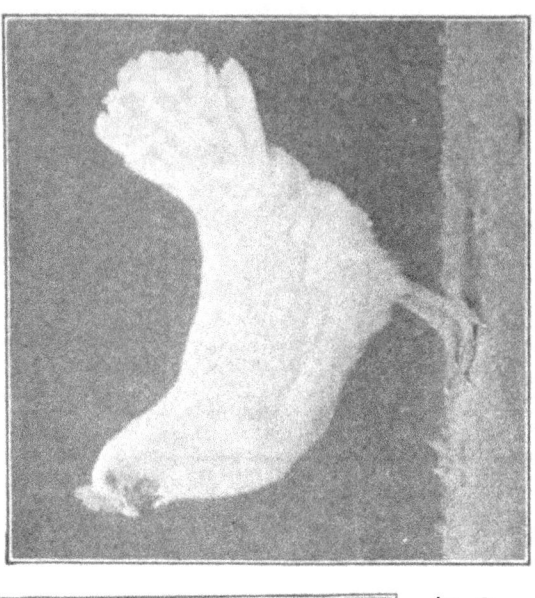

ROSE-COMB BROWN LEGHORN HEN
442 eggs in two years. (Utah Station.)

BELLE OF JERSEY
649 eggs in three years. (New Jersey Station.)

THREE CORNELL LONG-DISTANCE LAYERS

Cornell Supreme, 665 eggs; Lady Cornell, 648 eggs; Madam Cornell, 539 eggs in three years.

of the great cost of incubating and rearing the chicks. The first result would be a reduction of the number of market chickens, as with half the number hatched there would be half the number of surplus cockerels and half the number of hens sold. But poultry producers would find a better market for meat chickens, the production of which would develop into a more specialized industry.

LADY MACDUFF

Oregon Station hen C521. Photograph taken day after she laid her 303d egg. The world's greatest layer so far as authentic trap-nest records show. She laid 303 eggs in 12 months; 512 eggs in 24 months; 679 eggs in 36 months.

THE 303d EGG OF LADY MACDUFF

Weight of first year's eggs approximately 42 pounds. Her three years' production, 95 pounds. Weight of hen, 5 pounds.

LADY MACDUFF AND 10 DAUGHTERS
The daughters averaged 250 eggs in year.

Highest Annual Records.—The table on page 120 gives the highest official individual records. They may be considered world's records, so far as official reports have been published.* They were all secured at the Oregon Station.

*One exception should be made. Lady Showyou (p. 121) made her record of 281 eggs in her second year at Missouri.

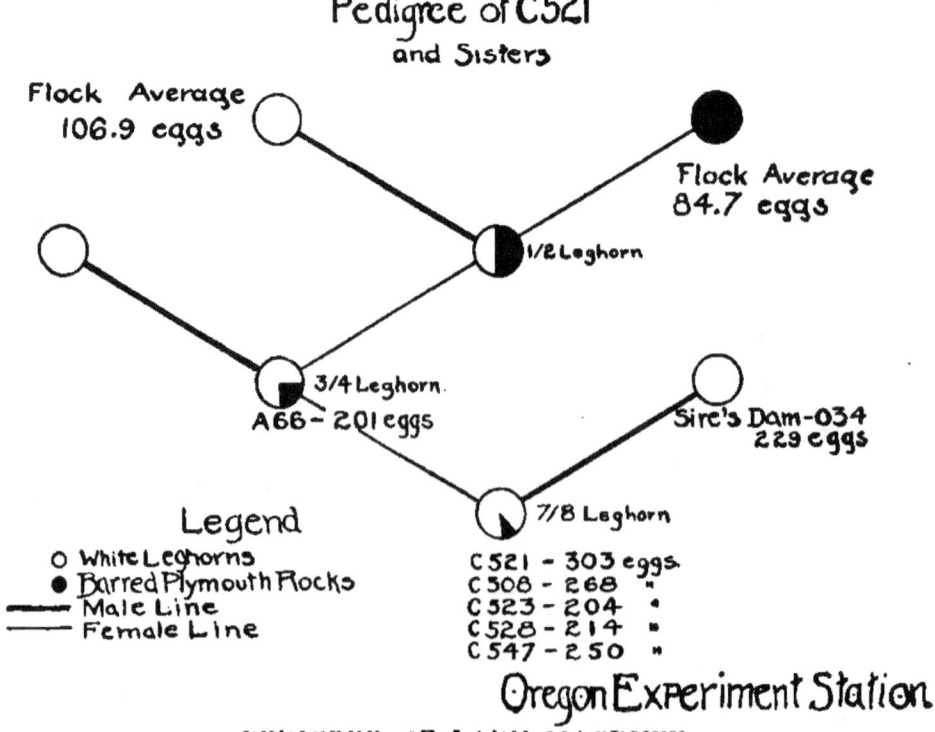

PEDIGREE OF LADY MACDUFF

SON OF LADY MACDUFF

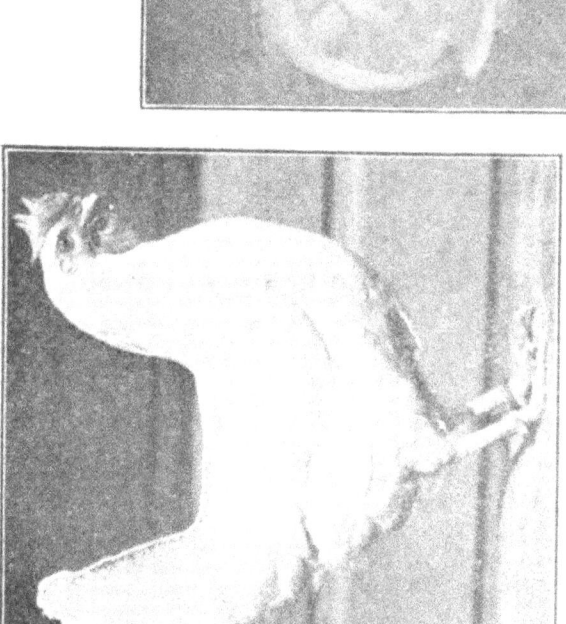

LADY MACDUFF

As she appeared in full plumage in her second year.

This trio illustrates type of new breed of general purpose fowls being developed at the Oregon Station. Female, about 4 pounds dressed.

DAUGHTER OF LADY MACDUFF
Laid 254 eggs in year.

120   POULTRY BREEDING AND MANAGEMENT

No. of eggs

First laying year (1912-13), Hen C521..................303
Second laying year (1912-13), Hen B42..................250
Third laying year (1913-14), Hen B222.................231
Fourth laying year (1913-14), Hen B14.................198
For two years (1912-14), Hen C521....................512
For three years (1912-15), Hen C547..................689
For four years (1911-15), Hen B42....................834
For five years (1910-15), Hen A27....................987

Type in Layers.—There are certain characteristics that are present in the good layer and absent in the poor—not always, but on the average.

Weight Correlated with Laying Capacity.—It has been found that within the breed or variety the heavier producers on the average are those of lighter weight. Sometimes some of the heavy hens are heavy producers, but this is not true of the average. At the Oregon Station a pen of 47 Plymouth Rock hens averaged 160.9 eggs. Separating them according to weight into three groups the following result was secured:

A66—MOTHER OF LADY MACDUFF
201 eggs first year.

	Number	Average weight	Production First year	Two years
Heavy	10	7 pounds	141.1	236.1
Medium	18	6 "	163.4	268.5
Light	19	5 "	173.7	293.5

The eleven heaviest layers,—those hens laying over 200 eggs each—averaged in weight 5¾ pounds.

LADY SHOWYOU

White Plymouth Rock hen, laid 281 eggs in the Missouri Contest in 1911-12. This record was made in her second year. Note wide-awake, active temperament. Weight 6 pounds.

It would be a serious mistake, however, to select year after year, the smallest individuals for breeding purposes without regard to other considerations. Vigor and health must always be uppermost. Continued selection of the smallest would, in the Leghorn breed, for example, finally evolve a Bantam type so far as weight is concerned. On the other hand, it is a mistake to pick out the nice large hens and the nice heavy males and save them for breeding, better send them to the pot.

Shape or Conformation.—Much importance cannot be attached to various theories regarding shape as indicative of laying qualities. The good layer, however, is usually medium to long in body, and rather deep and broad. These are relative terms and subject to breed differences. Em-

C543

As she appeared at end of first 12 months' laying. Her dam laid 402 eggs in two years.

OREGON STATION HEN C543

Laid 99 eggs in 100 days and 291 in first 12 months of laying. She is ⅝ White Leghorn and ⅜ Plymouth Rock blood. Photograph taken in her second year.

WHITE LEGHORN

286 eggs.

Highest record hen at the Missouri 1913-14 competition. Note nervous temperament. (Owner, O. E. Henning, Nebraska.)

A122 laid 259 eggs

A77 laid 214 eggs

A84 laid 44 eggs

A94 laid 20 eggs

THE HEAD INDICATING LAYING QUALITY

Masculinity is apparent in the head of the poor layers. In other words, the poor layer has a suggestion of a rooster head. The head is large, the comb coarse and the face fleshy. (Oregon Station.)

BREEDER OF POOR LAYERS
A94—20 eggs in a year. (Oregon Station.)

phasis is placed on these points as indicating digestive capacity, for the heavy layer must have good digestion. No great reliance, however, can be placed on shape of body as a method of identifying the good layer. There are good layers with short bodies and poor layers with long bodies. (See Page 93.) The truth is that hens have not been bred systematically for high egg laying long enough to fix or develop any particular type as it relates to shape.

The same can be said of the angle of the tail and the shape of the comb, though preference should be given to a rather large comb and a tail carried rather high. The head should be rather small, and leg bones not too large. In general make-up the fowl should not have what might be called a beefy build; rather, a trim, muscular build. A poor layer will usually have at the end of the laying year a better appearance than one that has made a heavy record. The hen at the

WHITE LEGHORN HEN
Laid 1 egg in a year. (Oregon Station.)

end of the year that is ragged in plumage and wrinkled in face—in other words, one that shows the effect of hard work—is more often the one that has been doing the laying. The hen that looks best at the end of the year, in the fall, is not usually the one that should be kept for breeding.

Hen Temperament.— The poultryman who is a close observer will

A GOOD LAYER FROM POOR LAYING STOCK

Laid 223 eggs. Dam laid 142; dam's dam, 74; sire's dam, 20. (Oregon Station.)

D33 laid 67 eggs—a good poser but poor actor.

D7 laid 7 eggs—good looker but poor layer.

POOR LAYERS FROM GOOD LAYING STOCK. (Oregon Station.)

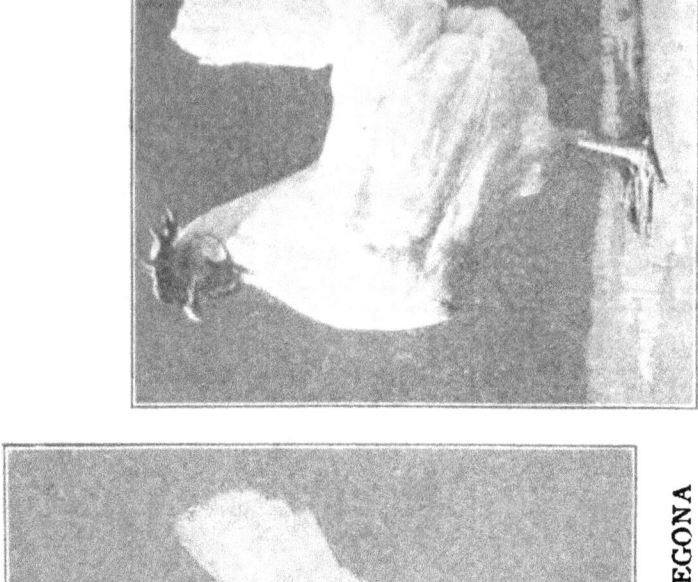

ANOTHER VIEW OF OREGONA

Oregon Station White Leghorn hen A27—987 eggs in 5 years. Note vigor and alertness. Medium length and depth of body. (Compare with photograph on page 108.)

UTAH STATION WHITE LEGHORN HEN

242 eggs in first year. Note short body, blocky build and high tail.

BARRED PLYMOUTH ROCK HEN

Laid 74 eggs in a year. Note beefy type.

find that temperament is correlated with egg-laying qualities. The good layer has an active, nervous temperament. She moves around quickly, and is "on the go" more than the poor layer. She doesn't pose well either in the exhibition coop or before the camera. She will be found scratching and hunting for food after the poor layer has gone to roost, and she will usually be at work early in the morning.

WHITE LEGHORN HEN C516

267 eggs in a year; 421 eggs in two years. Note high tail and erect comb. Mother of E248 (302 eggs) and other good layers. (Oregon Station.)

First Year's Production the Best.—Where fowls are kept under the same conditions each year and come to laying maturity in the fall, the production in the first or pullet year will exceed, on the average, the production of any subsequent year. Occasional individuals lay more the second than the first year, but this is exceptional. The production of a flock of fowls in the first and second years is shown

WHITE LEGHORN HEN E248
302 eggs in a year; dam C516; sire's dam, Oregona.

Laid 44 eggs.   Laid 79 eggs.

TWO POOR LAYERS

Note heavy, coarse build. (Oregon Station.)

Laid 104 eggs.   Laid 190 eggs.

UTAH STATION WYANDOTTES

Note the business appearance of the one and absence of it in the other.

PROBLEM OF HIGHER FECUNDITY

NEW ZEALAND WHITE LEGHORNS

Winners in the British Columbia laying contest 1913-14, averaged 221.7 eggs in 11 months.

WHITE WYANDOTTES

Averaged 208.5 eggs in the Storrs 1913-14 contest. Owned by Tom Barron, England. Note absence of the blocky type demanded by the "Standard."

on page 131. There were 43 fowls in the flock. Among the 21 making the best record for the two years, it is seen that only two of them laid less than the average of 153 in the first year. Among the 22 of the poorest layers for the second year, only five laid more than the average in the first year. If those laying less than 153 in the first year— 18 in number—had been killed off at the end of the first

130 POULTRY BREEDING AND MANAGEMENT

WHITE LEGHORNS

Averaged 208.8 eggs in the Storrs 1913-14 Competition. (Owned by F. F. Lincoln, Connecticut.)

WHITE LEGHORNS

Winning pen in the Panama-Pacific International egg-laying competition. (Owned by Oregon Agricultural College.)

year, the average for the two years of the 25 remaining would be 320.6 instead of 283.5 as the average of the whole flock. The 18 poorest averaged in the second year 92. The 25 best averaged 158 eggs in their second year.

The unprofitable hens in the first year are, therefore, on the average, unprofitable in the second year. Knowing

PROBLEM OF HIGHER FECUNDITY

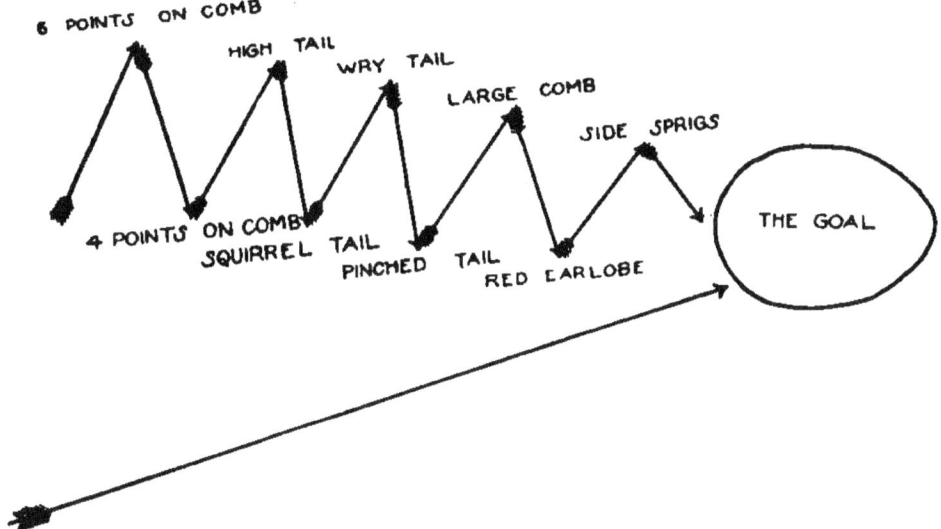

THE LONG AND THE SHORT WAY

In breeding for eggs slower progress is made in reaching the goal if various other points are bred for at the same time.

the production of each hen in the first year it is good business to kill off the poor layers at the end of that year. Further corroborative data are given on this subject in chapter on incubation.

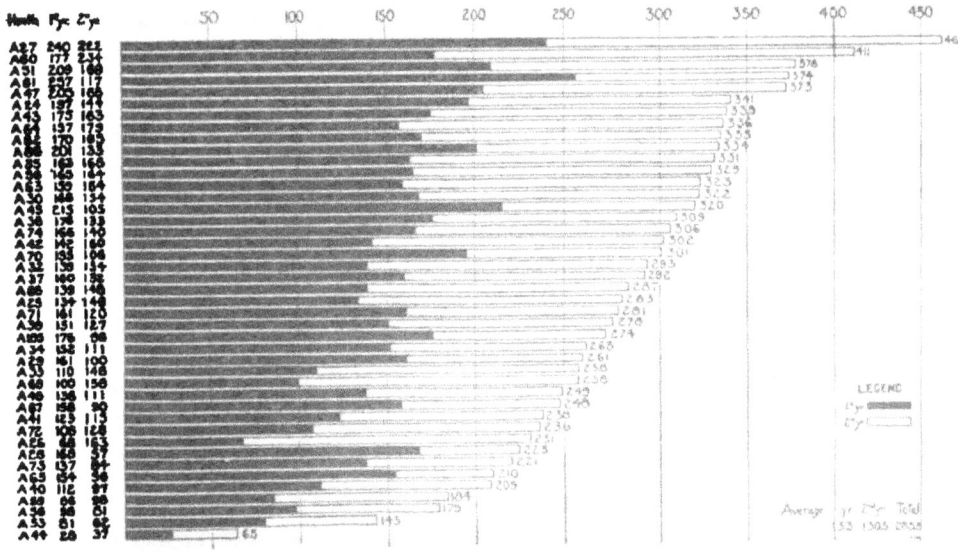

RECORD OF A FLOCK OF 43 FOWLS

At the Oregon Station for two years, showing that, on the average, the best layers in the first year maintain the distinction in the second.

Measure of Hen's Laying Capacity.—The egg record in the first three months' laying (November, December, January) will enable the poultryman to pick out the hens that will, on the average, prove to be the best layers during the year. If the poultryman will trapnest his fowls during those three months, he will find that the hens that lay 30 eggs or more in those months will lay during the year about 200 eggs. These should be kept for breeders. Those that lay less than 10 or 12 will,

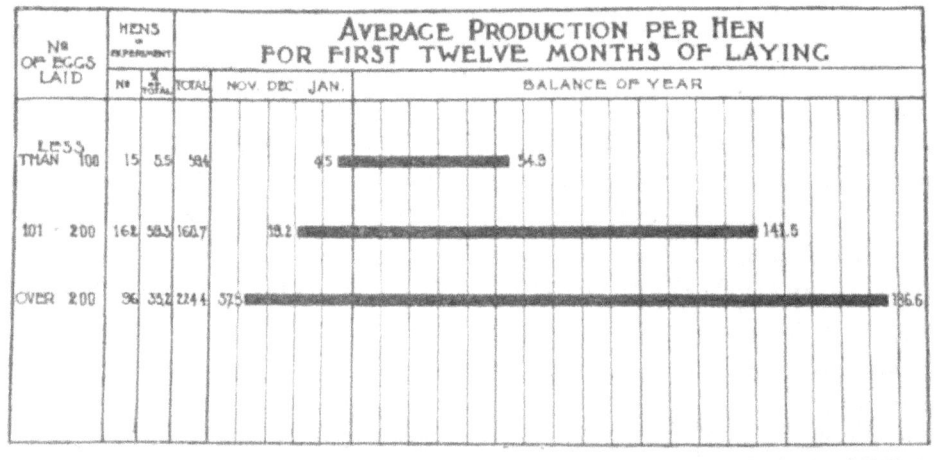

GOOD FALL AND WINTER PRODUCERS THE BEST LAYERS

on the average, prove to be unprofitable, and should be disposed of. Those that lay 20 may make a profit. The above conclusions are based on the Oregon experiments, which are shown above in detail.

Early Laying Maturity Characteristic.—Summarizing the Oregon Station records, it was found that pullets that began to lay under 200 days of age (approximately 6½ months), laid on the average about 200 eggs in the year. As this age advanced the number of eggs laid decreased. This is shown graphically on page 133. With or without the aid of a trapnest, the poultryman, by observing the date the first egg is laid, may pick out the pullets that will lay the best throughout the year. If the pullets have been hatched in March and April, and begin to lay in less than 200 days of age, they will prove to be, on the average, 200-egg layers. These should be marked and kept for breeders. Those that do not lay till 300 days of age should be killed off immediately or sold for market. There are exceptions, but on the average it works this way.

EGG-LAYING ORGANS

The egg-laying organs of the hen are the ovary and oviduct. Originally eggs were laid only that chicks might be

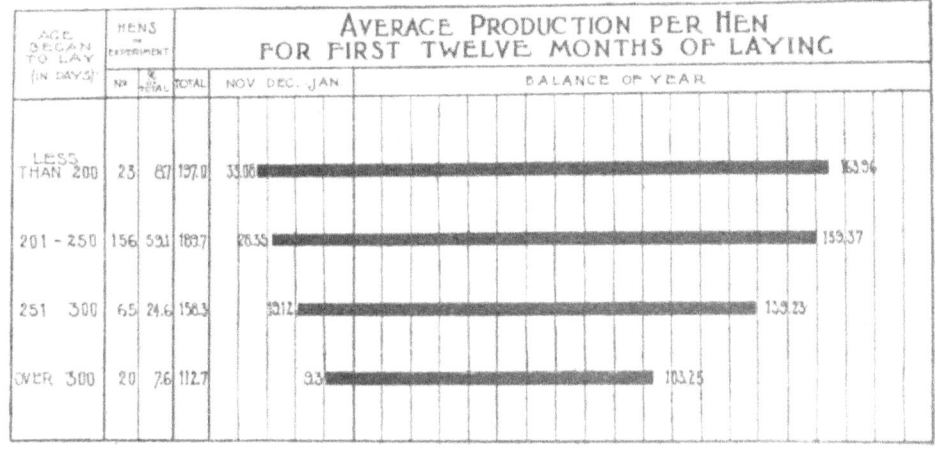

THE FIRST LAYERS THE BEST LAYERS

hatched from them; in other words, the purpose was reproduction. Later, when it was found that eggs were good to eat, egg-laying became a productive as well as reproductive process. They were to be used for food as well as for producing young, and it would be proper to call them productive organs as well as reproductive organs.

The *ovary* lies at the forward end of the kidney attached to the dorsal wall of the body cavity. The ova or eggs may be seen hanging like a bunch of grapes, though some are larger and some smaller than grapes. Some are so small as to be scarcely visible to the naked eye. From that they vary in size to the fully formed egg yolk. Each ovum or

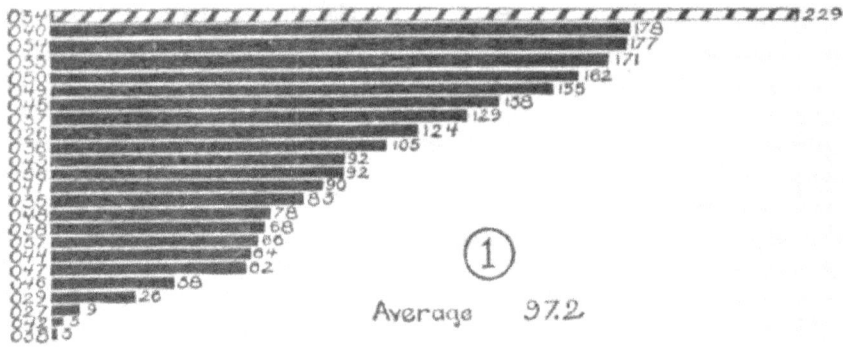

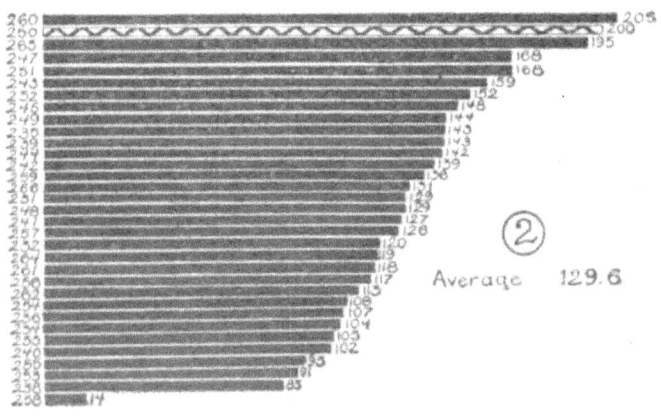

INHERITANCE OF EGG PRODUCTION, SHOWN DIAGRAMMATICALLY

These diagrams show the production of four flocks in four different years. Flock 4 is the pullet progeny of the best layers in flocks 1, 2 and 3. The hen numbers are given at the left and the eggs laid by each at the right. The dam is represented by the same line or symbol as the daughters, except in the case of the black lines, where the ancestry of pullets is not given. A son of 034 was sire of all pullets except the daughters of 250, this hen being inbred to her son.

PROBLEM OF HIGHER FECUNDITY

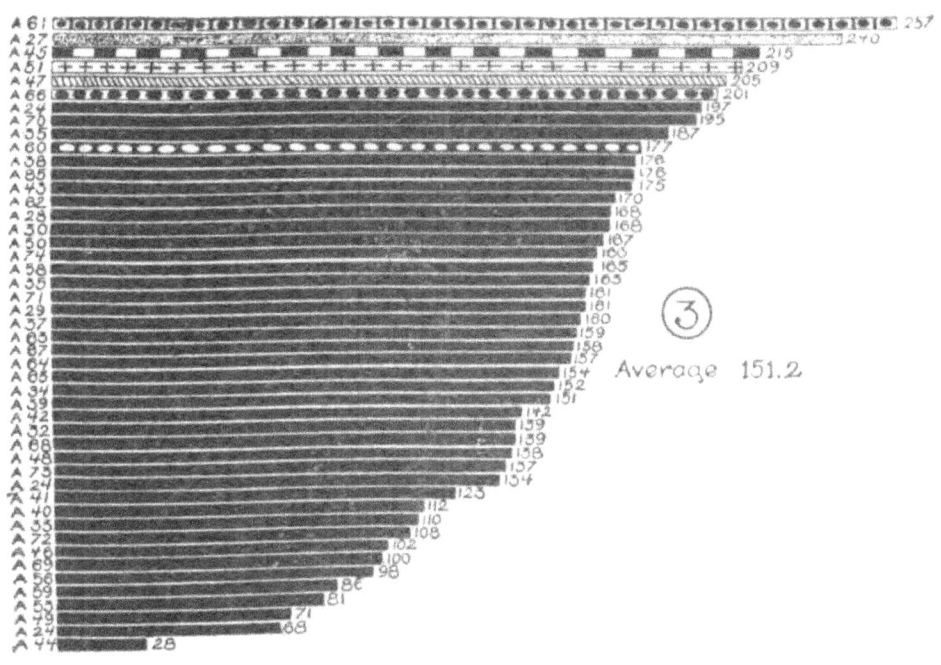

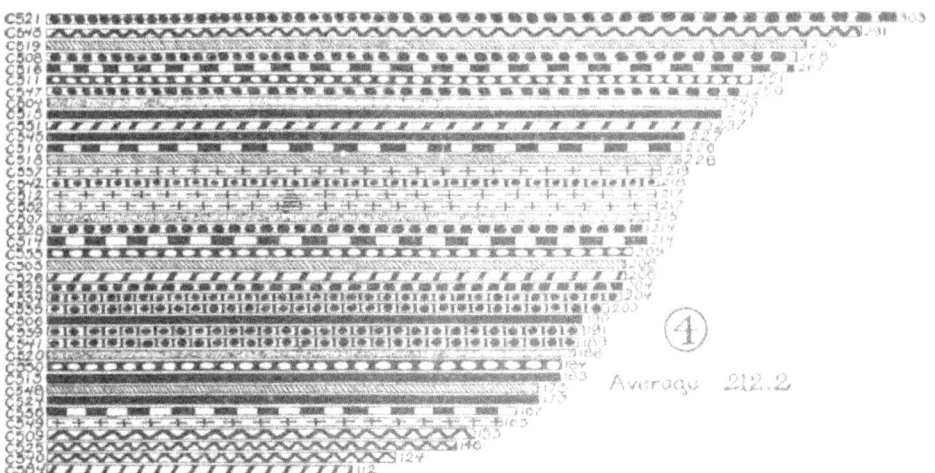

INHERITANCE OF EGG PRODUCTION, SHOWN DIAGRAMMATICALLY

These diagrams shows the production of four flocks in four different years. Flock 4 is the pullet progeny of the best layers in flocks 1, 2 and 3. The hen numbers are given at the left and the eggs laid by each at the right. The dam is represented by the same line or symbol as the daughters, except in the case of the black lines, where the ancestry of pullets is not given. A son of 034 was sire of all pullets except the daughters of 250, this hen being inbred to her son.

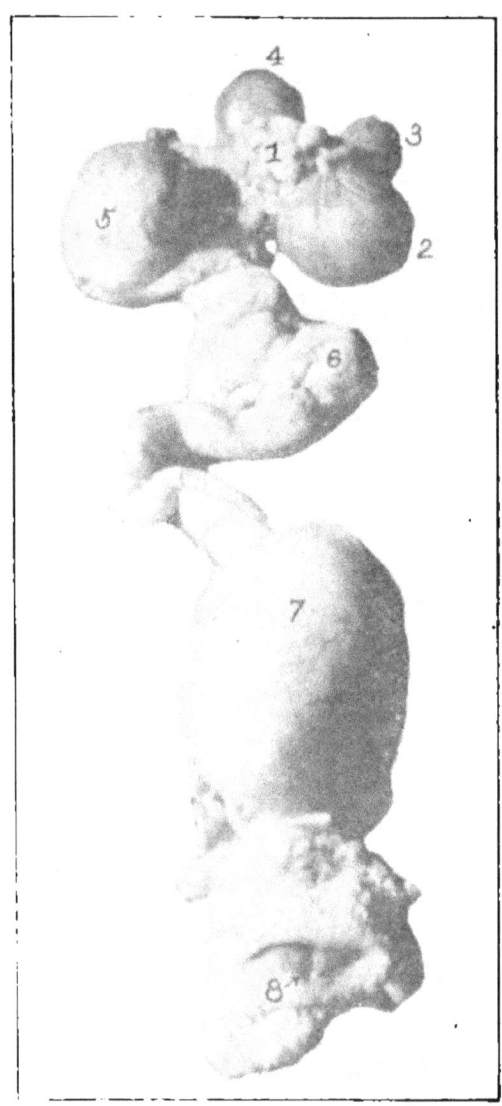

EGG ORGANS OF THE HEN

1. Ovary—Young follicles or eggs. 2, 3, 4—Larger follicles. 5—Ovum or egg yolk in upper part of oviduct. 6—Part of the oviduct where yolk receives the albumen. 7—Lower part of the oviduct showing complete egg in shell gland, ready to be laid. 8—Anus. (Oregon Experiment Station.)

egg is covered with a transparent sac. A normal hen has more than a thousand such eggs in the ovary. All the eggs that a hen may lay in a lifetime will be found in the ovary, in size from the smallest oöycote, visible only with the aid of a magnifying glass, to the mature yolk ready to burst from its sac. It has been found under normal egg-laying conditions that it requires about two weeks for the egg yolk to grow from the size of a pea to a full-sized yolk. The yolk is matured in the ovary. When matured it detaches itself and falls into the oviduct.

The rest of the egg is "made" in the *oviduct*. This is a large coiled tube, whitish in color, extending from a point just below the ovary to the cloaca. The albumin and shell are put on in the oviduct. This is accomplished more rapidly than is the development of the yolk. The perfect egg with its hard shell can be retained in the cloaca a short time, or several hours, before being laid.

In passing through the oviduct the egg travels about 24 inches. It is forced through this passage by contraction of the oviduct. As the yolk passes into and through the oviduct it becomes surrounded by albumin, and finally by the shell. The time occupied by the egg passing through the various sections of the oviduct is estimated by Kolliker as follows: In the upper two-thirds of oviduct, where albumin is formed, three hours; in the isthmus where the shell membrane is put on, three hours, and in the uterus for the formation of shell and laying, 12 to 24 hours.

Recent investigations by Pearl and Curtis would modify the above statement. (Maine Bulletin 206.) It was shown in their investigations that only 40% of the albumin was formed in the albumin portion of the oviduct; 10 to 20% was formed in the isthmus, or that portion where the membrane of the shell is made, and 30% to 40% of the total weight of albumin was added to the egg in the uterus, passing through the shell by osmosis.

CHAPTER VIII

SYSTEMS OF POULTRY FARMING

Various methods are followed in keeping poultry and it is well at the outset to form a clear conception of the business of poultry husbandry in its different aspects.

Mixed Husbandry.—The great bulk of the poultry and egg supply of the country is produced under a system of mixed husbandry. This type of farming constitutes the most promising field for increasing the production of poultry staples. Under present conditions poultry and eggs are produced at greater profit by the general farmer than by any other class of poultry-keepers. Poultry-keeping fits in well with about any system or type of farming. It is usually a side line, though sometimes it is the leading feature of the farm. The farmer may or may not specialize in poultry-keeping. Mixed husbandry may be carried on where the production of poultry and eggs is the leading feature and brings in the largest revenue of any branch of the farming operations.

Specialization.—This in poultry-keeping does not necessarily mean that the poultry-keeper must confine himself exclusively to poultry production. He may be a poultry specialist and grow the feed for his poultry and a large part of the food for the family. Specialization does not mean one-crop farming. The railroad business is a highly specialized business, but the railroad grows more than one kind of crop on its right-of-way. There is a freight crop and a passenger crop, and other crops, such as express, mail, etc. If the railroads were to specialize on passengers alone they would probably fail to make ends meet.

The best poultry specialization is that which makes the

SYSTEMS OF POULTRY FARMING 139

POULTRY KEEPING AND DAIRYING

A Petaluma ranch where cows and chickens use the same range. More than 5,000 laying hens and three dozen Jersey cows are kept on this farm, owned by T. B. Purvine. The hens' feeding troughs are fenced in from the cows, "the fence," as well as houses, being portable. (Two views.)

best use of land and secures the highest profit per head of fowls kept. Where mixed husbandry poultry-keeping is followed the cost of feed is comparatively low. The smaller the number of fowls kept per acre the lower the food cost will be. The ordinary by-products of other branches of farming and the waste grains will be sufficient to keep a small flock of fowls without any apparent feed cost.

Dairying and poultry-keeping is a good combination, the poultry furnishing a profitable market for the skim milk or butter milk. Possibly on the grain farms poultry may be kept at lowest cost. A combination of poultry with fruit growing may be successfully followed. Apart from the return in eggs and chickens the farm is benefited by the flock of poultry in the destruction of weed seeds and insects and in the manure furnished. The chickens often rid the fields of grasshoppers and other injurious insects. The manure from 50 fowls will maintain the fertility of an acre of land for the growth of crops. Poultry-keeping fits in well with a system of crop rotation.

There is undoubtedly an advantage also to the poultry themselves, under conditions obtaining in mixed husbandry. The large range is conducive to health and vigor in the fowls. There is no overcrowding of the land and the danger of soil contamination is largely eliminated. On the general farm the best conditions are available for the health and vigor of the stock and for low cost of production. Looking at it from the standpoint of the state or community, the development of this type of poultry farming offers the surest and quickest means of bringing production of eggs and poultry up to the demands of the consumer.

Examples.—Examples of profitable poultry farming under mixed husbandry conditions may be found in any county, but general farms on which poultry-keeping is conducted on rather a large scale or as the main feature of the farm, are not numerous. Little Compton, R. I., and Petaluma, Cal., are two districts referred to more generally than others where extensive specialized poultry farming prevails. The Petaluma district is largely given over to extensive poultry farming and examples of the same type of poultry farming may be found in Little Compton, but the latter could hardly be characterized as a district of exclusive or special poultry farmers.

A great many, if not the majority of the farms in this district, come more or less under the designation of mixed

SYSTEMS OF POULTRY FARMING

An Oregon fruit and poultry farm.—N. C. Jorgensen, owner.

A California poultry and fruit farm. Houses are portable.—H. A. George, owner.

COMBINATION OF FRUIT AND POULTRY

husbandry farms with poultry production as the leading feature. Under those conditions poultry-keeping has been a profitable business for more than half a century. The extensive system prevails; that is, there is wide range for the fowls. The colony system of housing is used. The hatching and rearing of the fowls is done by natural means almost universally, and in feeding, the general practice is to feed a moist mash in the morning. Farm crops are grown to a limited extent.

Exclusive Poultry-Keeping.—This may be defined as that type or kind of poultry-keeping that is carried on by the poultry-keeper as an exclusive business in itself and as an exclusive means of support and profit. All the feed for the fowls is purchased. There are sections in the United States where this type of poultry-keeping has been carried on successfully for years. It is true, however, that

1,000 PULLETS IN A PRUNE ORCHARD
Owners, Rev. M. C. Wire & Son, Newberg, Oregon.

a great many failures have resulted. This branch or type of poultry-keeping has been more or less uncertain in the past but there is not now so much excuse for failure because the available information on the subject is more reliable than formerly. Exclusive poultry-keeping must not be gone into by the novice without experience or he is almost certain to fail.

There are only certain districts or locations where poultry-keeping should be made a special or exclusive business.

Nearness to good markets and shipping points, reasonably cheap feed, low land cost, suitable soil and climatic conditions, are points that must not be overlooked when deciding whether to embark in this kind of poultry farming. These conditions being favorable, a man with the proper knowledge and experience, may successfully engage in exclusive poultry farming, and by giving special attention to the quality of the product. whether it be eggs and fowls

EGGS AND PEACHES FROM THE SAME GROUND
Chas. G. Weaver, Los Angeles.

for select market or breeding stock and eggs for hatching, he will be able to add considerably to his revenue and profits. It is here that a special or exclusive poultry-keeper has an advantage over the farmer with the small flock. He has enough eggs and poultry to make it a point to develop special markets.

Examples of Exclusive Farms.—Examples of exclusive poultry farming may be found in any state but there are few sections where any considerable area is given over to this type of farming. No doubt the district of Petaluma, Cal., offers the best opportunity for a study of this type of farming of possibly any district in the world. In this district a whole county is practically given over to poultry-keeping, and though it does not all come under the designation of exclusive farming, a large proportion of it does. Different types of poultry farms are here found varying from the intensely intensive to the very extensive, or in area from an acre of ground to over 500 acres. While intensive poultry-keeping is practiced the industry has been built up largely along extensive lines. There are also poultry farms of the mixed husbandry type, and combinations of poultry raising with dairying and with fruit growing are frequently seen.

Among the large, exclusive farms, there is much similarity in the methods or system followed. The colony house and free range system is almost universal either on the large or small farms. The success of the Petaluma district is doubtless largely due to this system. The house is one that may be easily moved by a team of horses. On some farms, though the houses are easily portable, they are allowed to remain for several years without moving. The usual size of the house is about 7 x 12 or 8 x 12 feet with gable roof. It is built on the box plan of construction, the frame consisting of runners, to which the cross-pieces are bolted at the ends, the plates and four rafters. The siding is nailed on vertically and serves to support the sides without studding. On some houses shingles are used, on some roofing paper and on others shakes. Floors are used in the houses on some of the best types of farms.

As understood in Petaluma the colony system is this: A colony of fowls on the large farms is usually 200 hens. For this colony two roosting houses and one laying house are provided, a section of the latter being used on many

FREE RANGE COLONY SYSTEM AT PETALUMA, CALIFORNIA

(1) A section of a 200-acre Petaluma poultry farm, showing prevailing type of house and feeding trough. (2) An 8,000-hen farm. Note colony houses dotted over the hills.

farms for storing feed. The different colonies are so widely separated that the grass is never eaten off the fields. The laying house is placed between the two roosting houses and is usually smaller than the roosting house. There is nothing in the roosting house except the perches, the whole space being used for roosting. With 100 hens in a house the roosting space per hen is about one square foot floor space. This, of course, is crowding them to the limit and

PETALUMA FARM OF 120 ACRES AND 6,000 HENS
A boy on horseback feeds the hens wheat in half an hour. Wm. Reardon, owner.

it is not the universal practice. In any less favorable climate it would not be advisable to do business on this basis, and even here it is very questionable whether better financial results would not be obtained if not more than 75 hens were kept in the house.

As to the system of feeding, the practice is to heavily feed a wet mash in the morning, with wheat in the afternoon. The feed troughs are fenced in if cattle or sheep are in the same field.

The White Leghorn breed is almost exclusively kept. Larger and less active breeds would not be suited in some particulars to the methods followed. It may not seem reasonable to say that a man with 5,000 hens on a farm

SYSTEMS OF POULTRY FARMING 147

of 100 acres or even 200 acres can get better results from his labor by colonizing his hens all over the farm than the man with 20 acres and 5,000 hens closer together, but the poultrymen with the large farms are undoubtedly handling the business at better profit than those on limited acreage near town. The saving of steps by building houses

CLEANING OUT THE HOUSES ON THE REARDON FARM

There are no dropping platforms. The roost perches are pushed out at one end of the house, the manure scraped out the door and thrown onto a sled, then lime is scattered on the floor.

close together does not necessarily lessen the labor or reduce the cost of producing a dozen eggs.

The important question is the maintenance of vigor and productive qualities in the fowls. Where the acreage is so limited that the ground is kept bare of vegetation the year around, is muddy in wet weather and hard and warm in dry weather, the conditions are more favorable for loss

A LARGE POULTRY FARM IN RHODE ISLAND
Where land unfit for cultivation is used.

of vigor and consequent lower production. On the wide range it should be understood that the work of feeding is much simplified. For instance, on a farm of 120 acres of which the author has knowledge, with 6,000 hens, the afternoon feeding was done by a boy of fourteen in half an hour. He jumped on a horse at one o'clock and made the rounds of all the scattered colonies in that time, doing the work of feeding wheat by opening a self-feeding bin. Under intensive conditions greater care must be exercised in the

2,000 HENS ON THREE ACRES
S. A. Bickford, near Los Angeles.

EXCLUSIVE POULTRY FARMING ON THE INTENSIVE SYSTEM
Suburban type of farms. (1) 500 hens on two acres, H. A. George, Petaluma. (2) 2,500 hens on four acres. C. G. Weaver, Los Angeles.

feeding. The problem of exercise is practically eliminated under the extensive free-range system. Under the intensive or yarded system more frequent feeding is necessary in order to induce exercise. Straw or other scratching litter must be furnished and the work of cleaning out the old straw and putting in new involves considerable labor and expense. The yards must be cultivated and possibly disinfected. Green feed must be furnished every day,

4,000 HENS ON FOUR ACRES
Swanson & Johnson, near Los Angeles.

involving expense both for the green stuff and for labor in feeding. The opening of gates through the yards is troublesome.

On some soils there is sufficient grit for the fowls on free range. This saves the expense of buying and feeding grit. Some animal feed will be found on free range. Under certain conditions there will be waste grain in stubble fields and weed seeds. These will lessen the feed bill. Again, as to equipment. There is considerable expense for fencing yarded fowls which is not necessary on free range. The increased cost for all these items under the intensive plan will largely offset the added labor cost of caring for fowls under the colony house system.

Suburban Poultry-Keeping.

As a result largely of the development of electric systems of railroads the keeping of poultry in the suburbs of cities is a type of poultry farming that has been making much headway in recent years. City people, with a love for the soil, build homes near electric lines a few miles from the city, and on an

THE INTENSIVE PLAN
A thousand hens eating green feed, southern California.

acre or two of ground are able to add to their income by keeping fowls. Where other members of the family can help with the work a profitable business may be done on an acre or two in the suburbs. With special care in the production of eggs and chickens of good quality accessible to markets, it will be possible to obtain good prices. The product may be delivered direct to the consumer either by private delivery or parcel post. If located on a good automobile road many eggs may be sold at good prices to passers-by.

Backyard Poultry-Keeping.—Chickens may be kept on a city lot at a profit. The waste food from the table of an average family, in addition to a little grain, will feed enough fowls to furnish the needed fresh eggs for the family. With good hens and careful attention to the houses and yards, a piece of ground 25 x 50 feet will accommodate enough hens to produce as many fresh eggs as the average family will consume, besides a considerable number of broilers. It must be understood, however, that the chickens will require daily attention throughout the year. The feeding must be done regularly and intelligently and the premises be kept clean and sanitary. It is not necessary that a chicken yard should be a disfigurement to the back premises; a chicken yard may be made a thing of beauty as well as profit on a town lot. It is not necessary to make the chicken yard a nuisance ground, or dumping-place for old shoes and tomato cans. Chickens do not thrive on such things. The spectacle of

BACKYARD EGG FARMING

Clarence Hogan, winner of a $100 prize in a poultry contest at Portland, Oregon. He was 13 years old and finishing the eighth grade in school. He fed the chickens, cleaned their houses, spaded the yards for them, weighed the food they ate and counted the eggs. In addition to all that he was an editor—editing a small paper and publishing it with the aid of a typewriter and a mimeograph. What can a boy not do?

a chicken yard made into a dumping-ground for rubbish and the chickens treated as scavengers is disgusting, and should not be tolerated in any community.

A nice flock of chickens properly cared for and housed,

BACKYARD EGG FARMING

Miss Ruth Hayes, winner of second prize, $50. A poultry yard may be made an attractive feature of the backyard.

and yarded in becoming style, may become an attractive feature of the vacant lot; and besides furnishing the daily fresh egg, will afford a mental diversion to some of the older people and a pleasure to the younger members of the family. Young boys and young girls of the town, lacking something to do, will find in a flock of chickens interest and instruction. The more ambitious youngster will find opportunity for a study of the problems that are of absorbing interest to all students of plant and animal breeding; for though the chicken is a chicken, it is subject to the same laws of heredity as are plants and livestock in general. Profit therefore may be realized in different ways from chickens on the city lot when given proper care and attention.

A Plan.—On a piece of ground 25 x 50 feet enough fowls may be kept to furnish the eggs needed for an average family. It should not be attempted, however, on unsuitable ground. It is not expected that the plan submitted can be followed under all conditions. The house suggested is 6 x 8 feet with a shed roof. The style or shape of the house may be changed to suit the tastes of the owner, but the amount of floor space or air space as provided for in the sketch should be available whatever the shape of the house may be, for 12 or 15 hens. At least a fourth of the side of the house should be open to admit fresh air. In cold sections a curtain of muslin may be hung over the opening at night, but this may not be necessary, as shown in the chapter on Housing. If there is not sufficient light a window may be put in the end of the house at a point where the maximum sunshine will be admitted. The illustration shows a little flat roof and it is covered with building paper. If shingles be used at least one-fourth pitch will be necessary. It will be best to face the house to the south to admit the sunshine, but it may be faced in any other direction to avoid strong wind, or for other reasons. (See illustration on page 156.)

A BACKYARD HOUSE

In this house 25 hens were kept for a year. The hens were never out of it. They laid an average of 188 eggs.

SYSTEMS OF POULTRY FARMING 155

The plan shows double yards each 12½ x 50 feet, less the space occupied by the house and the small lot at each corner. This means there will be "rotation of crops." The chickens may be rotated with the vegetable garden. The garden will be in one yard while the chickens are in the other. Chickens should not be kept on the same yard two years in succession. It is a mistake to crowd the yards beyond ability to keep them clean. The yard should be spaded frequently to prevent accumulation of droppings

ANOTHER BACKYARD SYSTEM

The original "Philo method" of close confinement. The system shown here entails a heavy labor cost when large numbers of fowls are kept.

and to keep the soil in better condition for the fowls to scratch in. This will furnish exercise for the fowls and help to keep the ground in a sanitary condition.

At the end of the house there is a small enclosure, 6 x 8 feet, which may be used for hatching the chicks if it is desired to rear them instead of to purchase mature pullets. Where it is possible to purchase the pullets in the fall from a breeder of good laying stock at a reasonable price, it will be more satisfactory to do so than to continue rearing the fowls on the small lot to reproduce the flock. The layers may be kept two years and then replaced with pullets.

The Crowing Rooster.—Where the method of buying the pullets is followed, the rooster is unnecessary. The hens will lay as well without him, and the objections of the neighbors to chickens on account of the early morning crowing will be overcome. If desired to keep a male, he may be discouraged from crowing by placing a board or hanging canvass over his perch at such a height as to prevent him stretching his neck. A rooster in crowing

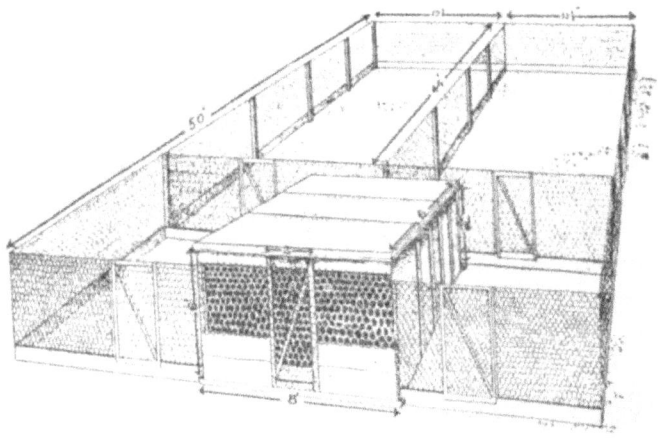

A PLAN FOR BACKYARD POULTRY KEEPING

raises his head at a considerable height, and if he cannot raise it the desired height there will be little crowing.

Fancy Poultry-Keeping.—Another type of poultry-keeping is that of breeding fancy or show fowls. It is not assumed that fancy or utility are not and cannot be combined, but there is a class of poultry-keepers whose chief business or profit is made in the production of fowls that excel in qualities demanded by the Standard of Perfection in the show bird. These breeders have been called fanciers probably because many of the points they breed for and make a profit on are matters of fancy rather than utility. Poultry fanciers have done a good deal in creating an interest in poultry-keeping through the medium of exhibitions. The breeding of fancy show specimens is a business in itself, requiring special fitness, and when by superior knowledge and skill in the mating of fowls the breeder is able to produce specimens so near perfection in exhibition points that others are willing to pay from $10 to $50 for single birds, he is not receiving more, probably, than reasonable recompense for the labor and skill expended in their production. Prices as high as $500 and more have been reported paid for single birds.

The standard followed by the fancier is not altogether based on points foreign to utility, but from the utility standpoint the fancier's standard emphasizes too highly many points of color and shape that have no correlation with useful qualities. So long as the leading poultry shows are judged according to this standard there will be a profitable business for the fancier, who has the necessary skill, in breeding prize-winning show specimens, even though he may not find a strong demand for his stock from the commercial poultry-keepers or farmers.

When the fancy standard is brought more in line with the utility viewpoint and show birds are judged more on a utility basis, the fancier will readjust his breeding practices and produce stock that is in demand not only in the show room but on the commercial farm. The business should then be more profitable and need not lose any of its fascination. If this is not done a double standard will be needed—one for the purely fancy, another for the utility.

A Financial Statement.—The following table gives the actual results secured on three different types of farms in the Willamette Valley, Ore., in one year. Similar types of farms could be found in any state of the Union.

Farm A represents that type of farming, mixed husbandry, which produces most of the poultry and eggs of the country, with the difference that the poultry branch of it is accentuated more strongly than on the average general farm. Farm B shows a combination of fruit growing and poultry raising on which poultry raising is of nearly equal importance with fruit growing in the average year. This farm shows heavy production for the acreage. Farm C represents the exclusive or special poultry farm type.

It is not assumed that these represent the best results obtainable on any one class of farms, but they are given because they show satisfactory results and at the same time give the necessary data for a study of different types.

Farm	A	B	C
No. acres owned	307	31	24
Value per acre	$93.48	$606	$1,000
Horses needed	7	2	1
No. of fowls kept	300	770	776
Estimate per fowl:			
Cost of feed	$1	$1.50	$1.25
Eggs per hen	125	140	125
Prices received:			
Highest	$0.47	$0.45	$0.65
Lowest	$0.18	$0.20	$0.27½

RECEIPTS

	A	B	C
Hay and field crops	$1,381.20	..	..
Animal husbandry	996	$ 10	..
Dairy	300	50	..
Orchard and garden	..	5,270	..
Value of poultry and eggs sold	839.30	2,715	$2,708
Total	$3,766.50	$8,045.30	$2,708
Net receipts	$2,077	$4,554.30	$520
Net gain	2,944	5,379.30	1,420
Per cent gain on investment	8.6	19.7	5.3

EXPENSES

Farm	A	B	C
Hired labor	$250	$1,000	$ 8
Operating expense	782.50	2,716	1,288
Total expense	1,689.50	3,491	2,188

CAPITAL

	A	B	C
Land	$28,700	$18,800	$24,000
Dwellings	1,500	2,000	500
Other buildings	500	4,000	800
Machinery and tools	200	500	50
Live stock	2,914	485	670
Feed and seed	100	50	100
Cash to run farm	75	1,000	400
Total	$33,989	$26,835	$26,520

CHAPTER IX

HOUSING OF POULTRY

Environment has much to do in the matter of getting eggs; that is, there is a close relationship between environment and egg yield. What is environment? The house or shelter is part of the environment of the hen. The kind of soil the hen ranges on or scratches in; the climatic conditions,—rain-fall, snow-fall, wind movement,—the size of yard or the amount of land, mode of getting feed, disturbing elements or noise that will cause fright, number of fowls in the flock, all these and many other things are part of the environment of the hen. If these conditions are favorable her environment is favorable for egg production.

Changes in environment have had a great deal to do with improvement in egg-laying qualities of fowls since the days of domestication. This point, however, is discussed under breeding. Taming the wild hen, putting her under more favorable environment as to shelter and feeding, is responsible for a large part of the increase in her egg-laying. Probably no other domestic animal is so sensitive to environment as is the laying hen. A dog running past the poultry yards and scaring the hens, and strangers going into the poultry houses, will cut down on the egg yield perceptibly, so sensitive is the hen to her environment. A slight disturbance in mode of living—a change from one yard or one house to another, a change in attendant, neglect or sudden change in the feeding—is reflected immediately in a lower egg yield.

If the business of egg production is harder or requires greater skill than the business of butter production, it is because of this one fact—the extreme sensitiveness of the hen to her environment. This lesson should be thoroughly remembered in embarking in the poultry business and especially in planning the poultry houses and yards.

Changes and Progress.—Probably in no other branch of poultry husbandry have ideas and methods changed more radically during the past ten years than in that relating to housing of poultry, and we are bound to say that the changes have been along the line of progress. There are still problems in poultry housing but they are being worked out surely. At the end of the last century dense ignorance of simple, elementary principles of housing might well characterize the state of knowledge on this subject. Not that there were not some isolated examples of proper housing; there were. But there was no general agreement among authorities as to what constituted some of the essential principles of housing. There were fierce contentions on the subject but lacking actual demonstration the contenders got nowhere. The change of methods has amounted to a revolution. No one agency has been responsible for this change, but probably without the demonstrations made at experiment stations there would not have been the progress that has been witnessed. The lesson has also been learned and taught by costly experience and experiment of practical poultry-keepers. Professor Gowell's work at the Maine Station deserves strong commendation. Not that he discovered any new thing in poultry housing but he put conflicting ideas to test and by actual demonstration brought poultry-keepers face to face with the problem.

Notwithstanding the great progress that has been made since the beginning of the century, the last word has not been said upon this subject. Progress has been made, though sometimes it has been made by going backward; and it would not be surprising if ten years from now some of the things now advocated are thrown on the scrap heap.

The difficulties come mostly from failure to understand the essential conditions of housing. There have been many costly experiments in the poultry business; there will be many more, no doubt, but few have been more costly than the experiment in housing. Fifteen years ago a great many poultry houses were built on the theory that warmth was the first essential of winter egg production. Houses were double-boarded and lined with sheeting paper. Even brick or cement houses were sometimes built. Some were built on the hot house plan with plenty of windows to admit the sunshine. The fallacy of this theory has been pretty well demonstrated and it is now fairly well understood that the first essential of winter egg production, as well as summer egg production, is the health and vitality of the fowls, not warm houses. Whatever kind of house best meets the conditions of health and vigor is the one that will give the most profitable egg production.

The following quotations taken from an early edition of Lewis Wright's "Poultry Book," will be of interest here. It describes conditions obtaining in 1813, a century ago, and it points a lesson. In speaking of Scotch fowls it is stated: "The hens are kept in as dry and warm a place in the house as possible; in cottages they generally, during the night, sit at no great distance from the fireplace; the consequence is that the farmer whose poultry are in the night time confined in places without a fire obtain no eggs; the poor people have them in abundance." Warmth is not inimical to egg laying. It is the attempt to make fowls warm without ventilation or artificial heat that is impossible. In the case of the Scotch cottagers a century ago the fowls were kept near a fireplace, and ventilation and dryness were furnished by the fireplace.

Natural and Artificial Conditions.—It is a well established law that domestication tends to enfeeblement. The fact that fowls have been under domestication two thousand years or more nullifies this law only in degree. But then, how is greater production secured under domestication than in the wild state? It is secured in spite of domestication. The fowl is placed under more favorable

THE EVOLUTION OF THE POULTRY HOUSE
The first and not the worst poultry house that was ever built.

conditions for production. She is furnished a regular and copious supply of food. Under the wild state the food supply is often precarious. That is one reason why in spite of domestication there is high production. But it requires the highest skill of the feeder and the breeder to offset this law of enfeeblement.

Houses and Vigor.—The hardest problem in poultry-keeping is how to maintain the health and vigor of the fowls. Housing has considerable to do with health and vigor. Ages ago, before domestication, chickens roosted in trees, and they still have a little of the wild nature.

Did you ever notice when the curfew of the poultry yard summons the fowls to their roost, that they usually go to bed on the branches of the trees if there is one near by? Not long ago the writer watched a flock of fine chickens "retire" for the night. The farmer had built good houses for the flock, but near the houses there was a giant oak tree decorated by nature with mistletoe. One after another the hens flew into this tree, hopping from one branch to another until some of them reached

A HOUSE WITH INSUFFICIENT VENTILATION
About the worst ever built.

the topmost branches, higher than the highest barn on the farm. It was interesting to see the chickens nestle down under the mistletoe for the night while the roosts in the poultry houses were vacant.

On another occasion the writer watched a flock of hens retire for the night where they had the choice of two houses. One was a sort of shed affair with one side about all open; it was a fresh-air house. The other was a closed house with a few small holes for ventilation. About nine out of every ten of the hens crowded into the open house, though they had originally been equally divided between the two houses. They preferred the fresh air house. If there had been a tree in the yard they probably would have preferred that to either of the houses.

On still another occasion the writer watched a flock of 1,500 Leghorns go to roost. Their houses were in a cherry orchard, but when dusk came on the cherry trees were covered with white fowls while the poultry houses were

AN UNSATISFACTORY POULTRY HOUSE

A house 800 feet long being torn down because it proved unsatisfactory. It had two bad points: (1) It was built on an incline, with no tight partitions, and there was a strong draught from one end to the other. (2) It was too closely built up in front.

practically deserted. The tree was the first but not the worst poultry house that was ever built.

There are times, of course, in severe storms when chickens prefer the shelter of a roof to roosting in a tree; but the lesson is, that fowls prefer the outdoor life, or the "simple life," and when put in close houses and compelled to live there under the mistaken notion that this is being good to them it is imposing conditions that will result in decreased vitality. Housing is really an artificial condition for chickens and it is a serious mistake in poultry-keeping to follow too closely artificial lines.

It should not be concluded, however, from what has been said above, that the best kind of housing for chickens is in the trees. It should not be inferred, either, that we should avoid all so-called artificial methods in poultry-keeping. While housing may be an artificial condition for fowls, nevertheless good housing is necessary if we wish to get the greatest profit. In a state of nature, fowls lay only during the breeding season, and it is necessary in order to get eggs during the winter season to surround the hen with conditions that are more or less artificial. If winter egg production is an artificial condition, then we must resort to artificial means to induce the fowl to lay in that season. The danger is that we are liable to forget the nature of the hen and compel her to live under conditions too highly artificial.

In a state of nature where the only purpose of egg production is reproduction, the hen does not lay all the year. The spring is the natural breeding season. A hen will lay in the spring under all sorts of conditions, but when eggs are 50 cents a dozen about the easiest way to make her lay is to chop her head off. Winter egg production is a fight against nature, against the wild nature of the hen. The troubles in housing poultry come from failing to recognize the nature of the hen, and in forcing the process of domestication too far. In a state of nature her wings answered the purpose of a house; she flew into a tree to get away from her natural enemies. She does it yet, but the enemy now is the man who builds houses that are as deadly as the prowling jackal of the jungle. Her wings were given her to escape her enemies. But we have no use for her wings. They cause the poultryman a good deal of trouble and expense. But the wings teach us a lesson. If the house doesn't suit, the hen will use her wings to get away from it. She prefers the tree to a poor house, and very often prefers the tree to any kind of a house.

The lesson is that the hen has still a little of the wild nature, and when we modernize her, when we put her under artificial modes of living, we are liable to get the same result that the nation got in putting the Indian under conditions of modern civilization. They cannot stand it, not until they have been bred to it by a long process of selection. We will get better results if we remember this fact and plan our houses accordingly. That is a condition that the hen imposes.

Purpose of Housing.—When we build houses for chickens we have in mind their health and comfort. We may be influenced in this by kindness for the fowls but more often by selfishness that looks for a full egg basket; that is, we usually build houses for fowls to make them lay more eggs. We may say, then, that the purpose of housing is to increase productiveness; poor housing will decrease it.

Location of Houses.—1. Soils.—Chickens will thrive on a great variety of soils, but certain kinds are more adapted than others to successful poultry-keeping. If possible, heavy clay soils should be avoided. They are hard to keep clean or sanitary. A rather light, porous soil is preferable. This is drier in wet weather and not as hard in dry weather as a heavy clay soil. A wet soil is colder than a dry one. The house should not be set in a mud puddle. That is as bad as setting it in a snow bank. It was Pasteur who tried to inoculate the chicken with anthrax, he did not succeed until he made the chicken stand in cold water. The temperature of the chicken was too high for the germ to develop, but after reducing the temperature by cold water on the feet and legs, he succeeded in inoculating it with the disease. This will show why certain diseases make greater headway where the chickens are kept in wet, muddy yards.

2. Drainage.—If the ground selected has not good natural drainage, provisions should be made either by under-drainage or by open ditches for carrying off the surplus water. The water should not be allowed to stand in the yard. Muddy feet mean muddy eggs. Dampness means catarrh, roup, rheumatism, tuberculosis, etc.

Sometimes the poultry house is put on a part of the farm that cannot be used for anything else and occasionally on a low sour soil too damp for the growth of cereals. Such a place should never be selected as a location for the poultry house. On the other hand it is possible to select land that is too dry. The nature of the soil undoubtedly has an influence on the growth and development of the chickens. Chickens make a more thrifty growth if kept on a soil that retains some moisture. A soil that becomes extremely dry and warm in the summer months is not the best. Hot sand soils, as well as clay soils that bake hard in the summer, do not afford good conditions for profitable poultry production.

The question as to how many fowls may be kept on an acre of ground depends a good deal on the nature of the soil. Many more chickens may be kept on soil that is rather light and porous than on heavy clay soil. Soil contamination will not have the same danger on the porous soil as on the clay soil.

3. Air Drainage is sometimes as important as soil drainage. Cold, moist air seeks the lower levels. It is better to locate the house and yards on higher levels, where there is some air movement to carry off the cold, damp air or prevent it becoming stagnant. Fowls should not, however, be exposed to high winds. You will notice that on windy days they mope around in sheltered corners or in houses.

This is not favorable to high egg production. Sufficient air drainage, without interfering with the comfort or activity of the hen, is the ideal condition. The houses may be built on the leeward side of an orchard or in the shelter of buildings. A wind-break of trees may be set out where necessary to provide shelter.

4. **Sunshine.**—If possible the houses and yards should be built where they will get the full benefit of the sunshine. Face them south unless the prevailing winds are from that direction. If the prevailing winds and storms come from the west or south the house may be faced east. It may even be necessary in cases to face the house north. In such cases windows may be put in the south side of the house to admit the sunshine. Sunshine is a germ destroyer and a better egg producer than red pepper or other condimental foods.

5. **Other Points.**—Other points that should be considered in locating the houses are (a) the convenience of the attendant, nearness to the feed and water supply will save in labor; (b) building the houses away from the other buildings will make it easier to keep the premises free from insect pests and rats.

Chickens that roost in trees have good health. They have constitutional vigor. They lay well except in severe weather. Their eggs are of good weight and hatch well. They very seldom have colds in coldest weather, while their sisters in warm houses will be running at the nostrils and have swelled head and ruffled feathers.

If the greatest problem is to maintain the health and vigor of the flock, and hens will maintain health and vigor without any houses, why then are houses needed for fowls? It is true that hens usually prefer to roost in the trees rather than in the houses.

Protection from Cold and Storms.—A good poultry house should afford protection from storms and severe weather. A little shelter from the winds and the storms will add to the comfort of the fowls and therefore to the egg yield. A cold wave or a sudden change to colder weather, means an immediate demand for increased fuel to keep up the heat of the body. In this case the fuel is the food that the hen eats, and the food that has been

A BOY WITH A "SAFE" HORSE AND "SPRING" WAGON
GATHERS THE EGGS
Reardon Farm, Petaluma.

going into the making of the eggs will be drawn upon for fuel purposes. It is the food that furnishes the heat of the body as well as the material for eggs. Any shelter therefore that protects fowls from storms or sudden changes in temperature is an incentive to egg production. Fowls maintain rugged health roosting in trees, but sudden and frequent changes in the weather to which they are subjected in the trees interfere with egg production.

Storms More Objectionable than Cold.—In most sections under most climatic conditions fowls will spend their nights in the trees in preference to the best poultry houses. Severe storms—driving snows or heavy rains—will send them into the house. It is the storm more than the cold that the hen objects to. In a scratching shed where the fowls are sheltered from the wind the hen will sing and keep busy all day with the temperature at zero; but hard winds, even on a summer day, will drive her from her picking and bug hunting in the fields to the leeward

A REAR VIEW OF THE MISSOURI HOUSE, SHOWING VENTILATION

side of the poultry house, where she will stand humped up and look as though she did not care whether school kept or not. In other words, you can keep the hen busy at a low temperature if she is sheltered from the winds and storms. Feathers will keep out cold but will not keep out wind.

On one occasion I watched several thousand hens at Petaluma hunting the shelter of fences during the middle of the day to escape the strong breeze that was blowing from the coast, though the day was otherwise nice and sunshiny. It is shelter rather than warmth that the house should furnish. If the proper shelter be furnished the hen will take care of the heating apparatus. All notions of the warm house should be abandoned, and a shelter built. This does not mean that warmth is injurious to health and vigor. Fowls maintain good health in the warm months. As the winter departs and the warm spring days come, the hen is at her best, her comb is the reddest. Nor does it mean that warmth is inimical to egg production. In the writer's experiments at the Utah Station a little artificial heat increased the egg yield (Bulletin 102), but it was apparently at the expense of vigor, for the fowls in the cold house weighed heavier than those in the warm house at the end of the winter. A proper system of artificial heat may stimulate egg production, but an economical, and safe system has not yet been found.

THE MISSOURI POULTRY STATION HOUSE
Ventilation is secured by slatted shutter and by opening windows.

A warmly built house cannot be made warm and comfortable without artificial heat in cold weather. Let us see. To make it warm the practice has been to double board it. It is tight boarded on each side of the studding, and under the boards there is building paper to make it airtight. Then glass windows are put in to give light and sunshine, and there must be double windows also. If the hens are to have fresh air there must be openings in the walls to let in fresh air and when fresh air is let in cold air comes in. A double wall of that kind, even without the ventilation, will not keep out the cold. It will keep it out a little longer than single walls; the changes in the temperature may not be felt so soon inside the warm house; but there is something else that makes houses cold. The thermometer does not always tell us how cold we feel. It does not always tell us how warm we feel. Dampness in the air in the summer intensifies the heat.

A house cannot be both warm and dry in cold weather without artificial heat. Why? During the day the sun strikes through the windows and raises the temperature inside. At night the heat will escape through the glass. The temperature will fall rapidly at night. There will be a great difference between night and day temperature. Now, warm air holds more moisture than cold air. During he day with a high temperature the air will be relatively lry; at night it will be relatively damp though the same amount of moisture may be in the air. If the temperature falls low enough the moisture in the air will condense on the walls. Moisture or frost on the walls indicates that the air in the room is as damp as it can be; in other words it is totally saturated. It also means that the house is cold, otherwise it would not condense. A warm house that is at the same time dry in cold weather without artificial heat, is an impossible proposition.

A damp house is a cold house. Chickens can stand cold air, but not cold damp air. By opening the windows during the day we keep down the temperature of the house and this keeps the air drier. Dampness on the walls indicates that the air is damp, not that the walls were damp. The moisture in the air condenses on the cold walls. Dampness is taken out of the air and put on the walls. The moisture was taken into the air during the day when the temperature went up, and at night as the temperature falls the capacity of the air to hold the moisture decreases and is condensed on the walls. With more ventilation this moisture would escape. With such a house the thing to do is to keep the temperature down during the day—equalize the temperature more between night and day by opening the doors or windows—and you will add to the comfort and health of the fowls.

A knowledge of this fact shows at once how futile it is to build double walls and put in double windows with the idea of making the house comfortable; it shows also how we may save about half the cost of the building. It means a saving in cost of building as well as in the condition of the fowls and in the egg yield.

We have not found that we can keep enough fowls in a house to keep its temperature up perceptibly. Horses and cattle keep a stable warm from the heat of their bodies, but we cannot crowd enough chickens into a poultry house to keep it at the same temperature of the horse and cattle barn and expect the fowls to maintain good health. This shows that the poultry house must have greater ventilation, must furnish more fresh air than is required in the cow barn or in the living room of human beings. There must be a more rapid change of air in the poultry house than in the horse stable, and if we keep exchanging the air rapidly enough by means of ventilators or open windows, there will, of course be little difference between the temperature of the house and the temperature outdoors. We must have plenty of ventilation, and we cannot expect to keep the house warm from the body heat of the fowls.

The attempt, therefore, to reproduce spring in winter, or to make the hen imagine in mid-winter that the natural laying and breeding season is upon her, by building warm houses, has not been a success. The failure is due to the wide range of temperature in them. The "warm" house is a hothouse during the day and a refrigerator at night, unless artificial heat be used. A so-called warmly built house is unreasonable without artificial heat, and artificial heat, in our present state of knowledge, is also unreasonable.

Ventilation.—A good poultry house should be well ventilated. Fowls require considerably more fresh air than farm animals. It has been estimated that a hen, in proportion to her weight, requires double the weight of oxygen that a man or a horse requires. The amount of air breathed per thousand pounds live weight of hens is given by King as 8,272 cubic feet in 24 hours; the requirements of a man being 2,833 and a cow 2,804 cubic feet.

Experiments at the Wye (England) Agricultural College showed that the health of the fowls bears a close relation to the amount of carbonic acid gas in the house. It was found that air in houses with proper ventilation should not contain to exceed nine volumes of carbonic acid gas to ten thousand volumes of air. The ordinary air in country districts contains about three parts. It would be impossible to arrange ventilation so that the air in the poultry house would be as pure as that of outdoors, but from the experiments quoted it is safe to assume that nine parts or under is not injurious.

When we speak of air being impure we think of the carbonic acid gas in it, yet this gas in itself is not poisonous. It is harmless, but associated with it is some other impurity, the exact nature of which is not known, that is poisonous. The presence, however, of carbonic acid gas is a sure indication that the air is impure.

Ventilation, therefore, resolves itself into a question of maintaining a low carbonic acid gas content in the poultry house. The more and larger the openings in the house, the more rapid the exchange of air and the lower the carbonic acid gas content. Reducing the number of fowls in the house decreases likewise the carbonic acid gas. Again, weather conditions will influence the amount. In cold weather there will be necessarily a more rapid exchange of air than in warm weather. Likewise high winds will decrease the amount by causing a greater circulation of air.

Methods of Ventilation.—Elaborate ventilation systems are not called for in poultry house construction. Usually the cost of such systems precludes their use by the practical poultry-keeper. For the larger portion of the United States the best condition of ventilation will be secured by leaving one side or one end of the house open.

Open-Front House.—In sections where the temperature gets no lower than zero the open-front furnishes the best method of ventilation. This much is beyond controversy. By open front is here meant a house with one side entirely open where the fowls roost practically in the open air. There is, however, a problem as to how low a temperature fowls will stand and maintain a satisfactory egg production. Ordinarily zero temperatures will not injure their health, because fowls roost in the trees all winter and maintain good health and vigor, but when the temperature reaches a certain degree of cold, egg production will be cut off. It is a question of the happy medium, and there is a lack of definite information as to the lowest temperature in the open-front house at which fowls may be profitably kept.

Fowls, however, that roost continuously in a cold house become hardy and stand low temperatures better than those that have been accustomed to warm quarters. Fresh air furnishes the necessary oxygen to keep up the heat of the body. The more pure air in the roosting room the better will the fowls be able to stand the cold. It may further be said that this method of ventilation will be better in any section, north or south; will give better results both in egg production and in health of fowls, than that type of ventilation which is found in houses that have more warmth but less pure air.

Curtain-front House.—The curtain-front house has also been used successfully in cold climates. Instead of leaving the front of the house open at all seasons adjustable curtains of muslin or burlap are used. They are closed at night as necessary during the winter. The objections to the adjustable curtain are; first, that it may not admit enough air, and second that it requires nightly attention in cold weather to adjust it. There is the further disadvantage that it may be closed or open when it should be the reverse, as the temperature or weather conditions in the evening do not always indicate what they may be before morning. In other words, it may be necessary to close the curtain in the evening but not necessary toward the morning, or the reverse. It will be better in cold climates to keep the curtain closed continuously during cold weather. When this is done there should be windows in the house to admit light and sunshine.

Adjustable Open-front House.—The wide open-front is impracticable in sections where the temperature gets much below zero. On the other hand, the curtain-front is impracticable for reasons stated. The curtain may be eliminated by using a modified form of open-front. By decreasing the size of the opening, the same purpose will be served as by covering the larger opening with canvas or burlap. In a section where the minimum temperature is zero, one side of the house may be practically all open. In such a climate sufficient ventilation for fifty fowls will be obtained by an opening 3 x 8 feet or 24 square feet of opening equal to about one-half square foot per fowl. In colder climates, with a temperature of 20 below zero, the opening may be decreased to a fourth or a fifth the size. A small opening in a cold climate will give better ventilation than a larger opening in a warm climate. In summer the opening should be larger than in winter.

Space Required per Fowl.—When poultrymen estimate the capacity of their poultry house on the basis of so many square feet of floor space per fowl, they are figuring on an insufficient basis. A few years ago it was enough to know that 8 or 10 square feet of floor space had given good results and half that poor results; therefore the fowls must have 8 to 10 square feet of floor space. That conclusion, however, was based on the warmly built, double-boarded, poorly ventilated house. Now the capacity has been increased by putting more openings and more ventilation into it, and the fowls do as well with 4 square feet of floor space

fresh air in the former than the latter. The capacity of the house, therefore, should be measured rather on a basis of purity of the air in it than by the amount of floor space.

In the Wye experiments it was found that in the winter months with the temperature near zero a house 7½ x 7½ feet and floor space of 1½ square feet per fowl gave good results when ventilated well. The carbonic acid gas varied in the tests from 4.8 to 8.5 parts by volume in 10,000, the latter result being on a still day. The air changed in this house about four times an hour. It was concluded that about 10 cubic feet of air space and 1½ square feet floor space in a house with proper ventilation are essential, and that the carbonic acid gas content should not exceed nine parts in 10,000 by volume.

The capacity, however, of any particular house must not be determined absolutely by a standard of air purity. It must also provide sufficient space for the activity or exercise of the hen. Where there is little or no snow, or where the chickens can be out of doors every day in the year, about

will apply to flocks of twenty hens or more. For smaller flocks a more liberal allowance of space should be made. Where the climate is such that the fowls will seek shelter part of the year rather than go out of doors in the yards and fields, 4 to 5 square feet per fowl should be provided. The house should be built high enough for a man to work in without bumping his head. The height will allow sufficient air space for the fowls.

The Final Test of a House.—The egg yield is the best test of the merits of a poultry house. The completeness of the egg records in different houses may well form the basis for a study of the relative merits of different kinds of houses. A profitable study may be made of the housing used where high egg records have been secured.

The Australian laying competitions conducted at the Hawkesbury Agricultural College for a number of years have produced very high records. The houses used were small, being 11 feet long and 6 feet wide, divided into two

A COLONY HOUSE WITH CURTAIN WINDOWS

At the Utah Station the temperature reached 12° below zero and the Leghorns shown in the picture escaped without injury to their combs.

pens for six fowls each. This is equal to 5½ square feet floor space. The outside yards were 87 x 17 feet. During the winter the front of the house was closed up, and there were wire openings at the bottom of the back of the house and at the top. The fowls had more yard space than is usual under intensive methods. Here we have a small house with open front in which records averaging over 200 eggs per fowl were secured. The climate is mild there, somewhat similar to some of the Southern states and to certain portions of the Pacific coast.

At the Utah Station excellent records were secured under more intensive conditions. The house was 100 feet long and 10 feet wide, divided into pens 5 x 7 feet, leaving an alley at the back. The outside runs were 5 x 40. There was a glass window in each pen which was opened during

A SCRATCHING SHED

This shed is an advantage where house room is limited. The ventilation in this house is secured through the door, which is covered with poultry netting only, and between the plate and roof at the rear.

the day except during severe cold weather. Here with 7 square feet floor space and 40 feet yard space, one pen averaged 201 eggs, and individual records as high as 241 were secured. At the Oregon Station the portable colony house has been used exclusively. This house is 8 x 12 feet with open front. A pen average of 212 eggs from 40 fowls was secured, with an individual record of 303. The yard space was about 150 square feet per fowl, the house being moved once during the year on to another yard of equal size.

In the Missouri laying competition, a well-ventilated, small house was used. In this house Lady Showyou made her remarkable record. The Storrs laying competition house is also a small, separate house. In this house Tom Barron's Leghorns and Wyandottes made their great records. The highest record secured at Cornell University, that of Lady Cornell, was made in a small house. At the Ontario Agricultural College the best records were made in a small house.

It is thus seen that good egg yields have been secured in houses of different construction. Namely, in long, con-

A CHEAP SHED

In this shed a pen of two-year-old Leghorns were housed for a year. One of them laid over 200 eggs. Lowest temperature about zero. (Oregon Station.)

tinuous houses, and in small colony houses of different types. All of the houses, however, have been either open-front or curtain-front. As between the continuous or long house and the small or colony house, while good records have been secured in both, most of the good records, and all of the high records, have come from colony or small, separate houses.

Hatching Quality of Eggs as Affected by Housing.— The kind of house and the conditions of housing have a marked influence on the fertility of the eggs and their hatching quality. The hatching quality of the eggs is even more dependent upon good housing conditions than is the egg yield. A system of housing may be, in a measure, successful so far as egg yield goes, but it may be a failure when it comes to securing eggs of good fertility and that will produce chicks of good vitality. The poultryman may be successful in getting eggs, but fail in getting chicks and in the business of reproducing his flock. Under some systems of close confinement in houses he may get a satisfactory egg yield if he puts into the house fowls of strong vitality,

COLONY HOUSES ON THE ALMY POULTRY FARM, LITTLE COMPTON, RHODE ISLAND

but the chances are that the breeding or hatching quality of the eggs will not be as good as where the fowls have wide range.

It is practically impossible to make a permanent success of poultry-keeping where the fowls are confined in houses all the time if the eggs from the same stock are used for breeding purposes. If it is desirable to confine laying stock in such houses the layers should be produced by breeding stock kept under more or less free-range conditions. The natural tendency to enfeeblement under domestication asserts itself more clearly and strongly in the breeding qualities of the fowls than in their laying qualities.

The Floor.—Floors are not always necessary nor desirable in poultry houses. Where the ground is inclined to be damp, a floor will be an advantage, but where it is well drained and porous there need be no floor. Fowls prefer to scratch on the ground rather than on the floor. A wooden floor gives protection for rats underneath, and for this reason a cement floor is preferable. Where an earth floor is used it should be higher than the ground outside the house to prevent water running in. It is a good plan to fill the floor with 6 or 8 inches of clean, coarse sand and once a year or oftener take off part of this and replace it with clean sand. This will keep the floor comparatively clean and sanitary.

The Roof.—There are three types of roofs generally used, namely, the shed roof, the gable roof, and the combination roof.

Shed Roof.—Practically the same amount of material will be required for each style of roof, if the ground plan and the air space content of the house are the same. There will be a little less labor in constructing the shed roof than the others. The type of roof used will depend mainly on the width of the house and the pitch of the roof. If shingles are used, a comparatively steep roof must be

PIANO BOXES UTILIZED FOR HEN HOUSES

Two such boxes used in making this house.

made, about one-third pitch. If roofing material is used instead of shingles, a roof nearly flat may be made. Another advantage of the shed roof is that the rain drains off at the back of the house. This will obviate much of the mud, caused by the dripping from the roof, without the use of eave troughs.

A shed roof house may be made as much as 14 feet wide. For a house of that width to be shingled a gable roof of one-third pitch should be used. In most sections the cost of the roof will be lessened by making it comparatively flat and covering it with roofing material. In sections where shingles are cheaper there will be little difference in the cost between the shingles and a good quality of roofing paper. The cost of laying the shingles, however, is greater than for laying the roofing paper. Heavy tar or rubber roofing should be used. A good shingle roof is the most durable type of roof, though roofing papers are now made of good quality. It takes 750 shingles for each square of 100 square feet, laid 5 inches to the weather. A man can lay about 3,000 to 6,000 shingles a day, the latter by an expert shingler working on a large roof.

Foundation and Floor.—Where a permanent or stationary house is to be built, it will pay to put in a good foundation. Either brick, stone or concrete may be used, depending mainly on the cheapness of these materials.

CHAPTER X

KIND OF HOUSE TO BUILD

Poultry houses may be divided into two classes: 1, portable houses; 2, stationary houses.

Portable House.—The portable house is used where the colony system prevails. Much of the trouble from diseases comes from keeping the chickens on the same ground year after year. By keeping them on clean ground, which is possible with portable houses, they are under natural and hygienic conditions. This system, moreover, fits in with a system of crop rotation on the farm. About fifty fowls to the acre will keep the land in high fertility. Besides, the chickens will find a considerable portion of their food in the waste grain and weed seeds, grasshoppers and other insects. They often rid the farm of grasshoppers and other injurious insects, thus saving valuable crops. Another important advantage of the colony house system is the fact that the fowls are more active when they have the liberty of fresh fields than when confined in yards. Finally, with the colony system there is no expense for fencing. Where fowls are kept on an extensive scale this system is undoubtedly the best.

While the advantages of the colony portable house are more apparent in sections where there is little or no snowfall in winter, they may also be used where the snow covers the ground in northern sections several months during the winter. In such cases it is usually advisable to pull the houses near together to avoid the inconvenience of the deep

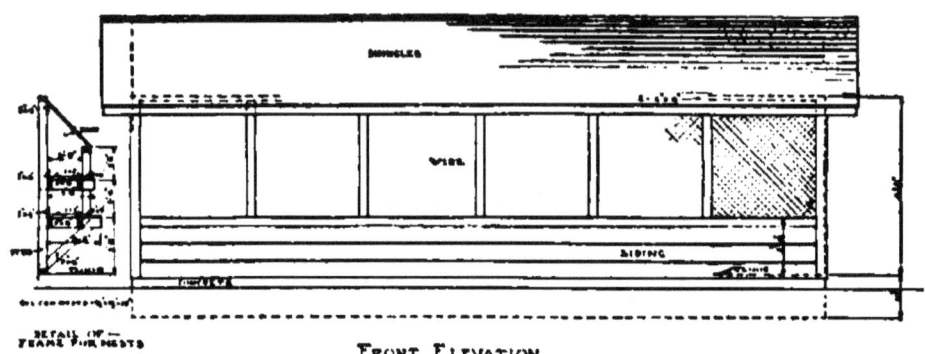
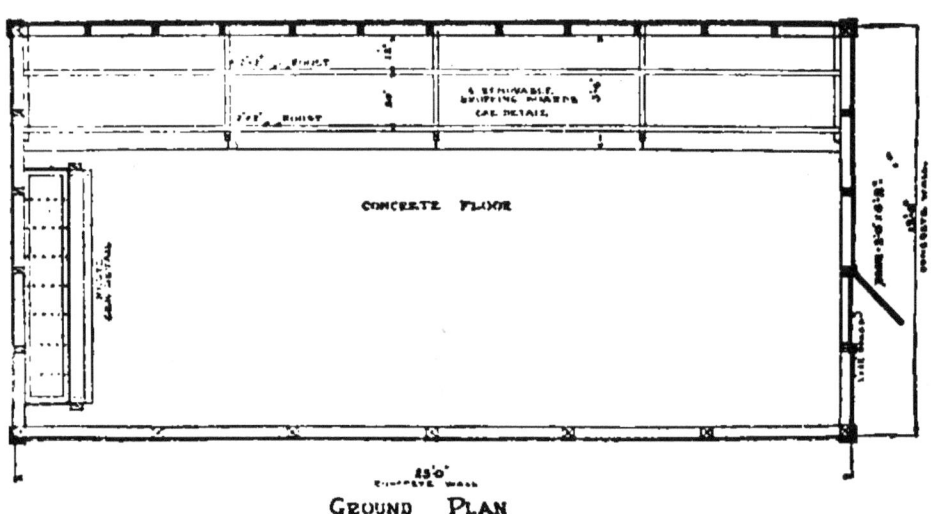
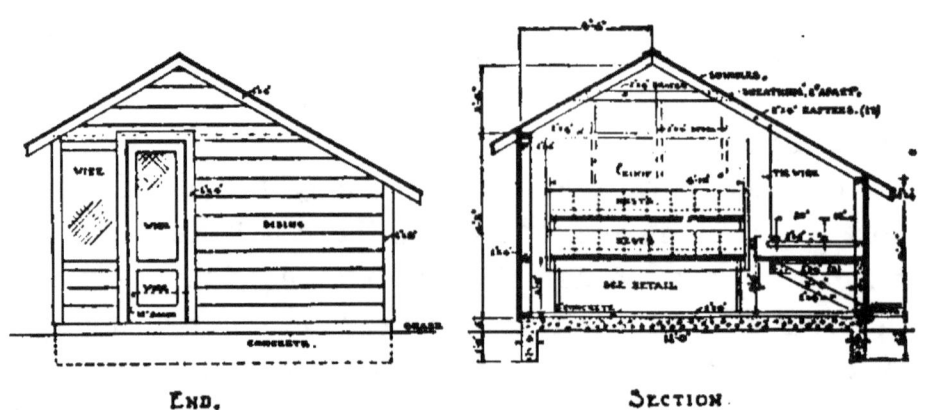

STATIONARY 100-HEN HOUSE. (OREGON STATION.)

snows. When the snow covers the ground and the fowls have not the use of the range, being practically confined to the house, they will do just as well when the houses are brought together.

Stationary House.—Where little land is available, stationary houses may be used. The portable house, shown on page 193 may serve as a stationary house for a small flock. For larger flocks either a long, continuous house may be used, or small, separate houses. The separate houses may be placed in a row 40 feet apart. By this arrangement

A CURTAIN-FRONT HOUSE

Built by A. F. Hunter, at Abington, Mass. The curtains are shown in the second pen. This is a modification of the scratching shed-house of which Mr. Hunter was the originator.

the yards may all be on one side of the house, and one can walk or drive a team on the other side from one end to the other without opening of gates. Another advantage of this arrangement is this: by having every other yard vacant the trouble from males fighting through the fence is avoided. Another advantage is that there is less danger of contagious diseases spreading from one flock to another than in the continuous house; every flock is practically isolated from the other.

Keeping large flocks in a long, stationary house requires less time for the feeding and caring for the fowls than in portable houses scattered widely apart over the farm. It does not necessarily follow, however, that the advantages in this regard are all in favor of the long, stationary house. The profit in the business does not hinge altogether or mainly on the convenience of the attendant or on the amount of time necessary to do the actual work in feeding and caring for the fowls. The final result must hinge rather on the results or on the returns in egg yield from a given amount of labor.

The portable house and free-range system is most conducive to health and vigor in the stock, and in the long run the financial results must be decided in favor of the system most favorable to vigor. A man may care for more fowls in a long, stationary house than under the free-range colony house system, but in a series of years will there be greater return in egg yield from his labor than from the labor of the man who keeps his fowls under the extensive free-range system? The greater risk from loss of vigor, from death, from contagious disease, from lower fertility of eggs, and greater mortality in the chicks makes it certain that in ten years, more or less, there will be a greater return from the labor on the colony free-range farm.

It is possible that under certain conditions of soil and climate the long, stationary house system may be successful for a long term, such as in sections of maximum sunshine and on porous soils. The sunshine will ward off many bacterial diseases which would be more common where there is not very much sunshine. Again, in a very porous soil, soil contamination has not the same dangers as in heavy clay soils. The poultryman who uses stationary houses and follows the intensive system must utilize to the utmost the assistance of the sunshine in warding off diseases which in many sections of the country have followed in the wake of intensive poultry culture.

Cultivation of Yards.—If the intensive system be used it is imperative that it include a system of cultivation or crop rotation. To allow the fowls to run in large numbers on the same ground, year after year, without any cultivation and growth of crops will result in certain failure. The cultivation and cropping of the yards will keep them in good condition. The crops will use up the manure and lessen the danger from spreading of disease. The cultivation also keeps the surface of the soil loose; unless cultivated, some soils of a clayey nature will, from continuous use, become hard and packed. The expense of building the extra fence for the double yards will be offset by the value of the crops that may be grown on the vacant yards. Cultivation has a double purpose; first, it cleans the yards; second, it offers the fowls more exercise. Whether it will require cropping every year or every other year, or twice a year, will depend first on the nature of the soil; second on climatic conditions, and third on the number of fowls kept on the ground. The control of tuberculosis is rendered comparatively easy by crop rotation and keeping the fowls off the ground for six months each year.

Capacity of an Acre.—A light, porous soil has a greater capacity for fowls than a heavy soil or a damp soil. At the Oregon Station on clay soil it was found that the day droppings from 200 laying hens on an acre in four years made the soil too rich for the successful growth of cereal crops where cropping the ground was done every other year. The night droppings were put onto other land. If the soil contains too much manure for the crops it is safe to assume that it is not in the best condition for poultry. Sooner or later it is bound to show not only a failure of grain crops but failure of poultry crops. For a permanent system under average conditions of soil and climate the following points are suggested for consideration.

1. Maximum number of fowls per acre: 100 laying hens.

2. Disposing of the night droppings on other land.

3. Dividing the ground into at least two divisions or yards, and growing a crop on each yard at least every other year. In sections where crops may be grown every year the maximum number of fowls may be increased.

4. Growing crops that will use up the maximum amount of manure.

5. Keeping the ground vacant at least six months in the year.

6. Thorough underdrainage, where necessary, to carry off surplus water.

The above points are suggested as worthy of careful attention where more or less intensive poultry-keeping is to be followed and where the location is expected to be a permanent one. It cannot be assumed that they will be applicable or practicable under all conditions of soil and climate. But under average conditions of soil and climate they afford a safe basis of estimating the capacity of an acre in a permanent system of poultry culture. It is not assumed that as many as 500 hens may not be profitably kept on an acre for a few years under favorable conditions. It has been done, but it is a different matter when it is planned to make a permanent business of it.

Crops to Grow.—Different kinds of crops or vegetables may be grown on the vacant yards. Green food may be grown for the fowls, or vegetables may be grown for the family. The droppings of the fowls will keep the soil in a very productive condition. If it is not desired to use the yards for garden purposes, such crops as vetch, clover, kale, rye, etc., may be grown. Where it grows well, clover may be sown early in the spring and the chickens turned on it in the fall. Vetch sown in the fall will furnish a great quantity of excellent green food in the spring and summer. Where it thrives, probably no other forage plant will furnish more green food per acre than the thousand-headed kale. If planted early in the spring, it will furnish a great quantity of green food in the fall and following winter. Rye sown in the fall will make considerable green food in the following spring and summer.

NOTE.—In a personal letter to the writer Edward Brown says: "For those who keep their fowls within restricted areas, I believe we shall have to come to a four-course rotation, fowls being one part to three others, by which is meant, supposing we have four acres of land divided, the fowls shall occupy one acre only each year and no more, the three vacant lots being cultivated. In some cases the three-course rotation has been tried, but that does not seem to get rid of the manure completely. However, it is a question of experiment and therefore your observations are very important."

Portable Colony House.—A good size for a colony house is 8 x 12 feet. A team of horses will pull a house of this size and it will accommodate from 30 to 50 fowls. Thirty to 36 fowls will be enough in northern states, where the fowls have not the liberty of outdoors all the time. This house is built on runners and may be moved several times a year. It will cost to build, about $15 for lumber, $5 for hardware and paint, and $10 for carpenter work, the cost varying in different localities as the prices of material vary.

On page 177 is shown the kind of house used at the Oregon

Station up to the fall of 1913. If the proof of the pudding is in the eating, this house has been satisfactory. An average of over 200 eggs per hen has been secured in this house with a flock of 40 hens, and it was in this house that hen C521 made a record of 303 eggs in a year.

Improved Oregon Station House.—This house, however, has been modified with the idea of furnishing a still more copious supply of fresh air. A study of conditions led to the opinion that the exchange of air at night in the roosting end of the house was not rapid enough for the number of fowls in the house. For thirty fowls the ventilation is ample, but for forty or fifty it was decided that the fowls were too close together to avoid re-breathing the exhaled impure air. It had been noted that several of the highest record hens at the Station had roosted close to the door on a step up to the trapnests. Hen C543, with a record of 291 eggs; C508, with record of 268; A122, with record of 259, and a number of other high-record hens had formed this habit of roosting at the open door instead of back among the other fowls on the perches. This was roosting practically in the open air so far as fresh air was concerned, and it might lead to the inquiry as to whether fresh air is not, in itself, a good egg producer.

The improved colony house is shown on page 178. In the cooler parts of the north or where the temperature gets down to zero and snow covers the ground two or three months of the year, and for 30 to 35 fowls, the house with the end open instead of the side is probably preferable, because where the temperature is lower there will be naturally a more rapid exchange of air. In warmer sections the house with the side open instead of the end is to be preferred. In this house single walls are used made of rustic siding. Trapnests are placed under the dropping platform. Nests may be placed at the end wall of the house, in which case the roosts and platform should be lowered 10 or 12 inches. A three-quarter inch hole is bored through the ends of the runners. Bolts are placed through these holes and a chain attached for moving the house. A team of horses pulls the house. No curtains are used. The opening and door are covered with 1-inch mesh wire. The dimensions and bill of lumber and hardware follow:

BILL OF MATERIALS FOR PORTABLE COLONY HOUSE,

8 FT. x 12 FT.

Lumber

- 2 3x6 14 feet long runners.
- 2 4x4 8 feet long sills.
- 5 2x4 8 feet long sills.
- 14 2x3 5 feet long studs.
- 4 2x3 7 feet long studs.
- 2 2x3 8 feet long studs.
- 3 2x3 12 feet long plates.
- 14 2x3 6 feet long rafters.
- 8 2x3 12 feet long nest frames, etc.
- 2 2x2 12 feet long roosts.
- 3 2x2 3 feet long roost supports.
- 175 board feet 8 inch ship lap for flooring and dropping boards.
- 125 board feet 6 inch roosting boards and slats for dropping boards.
- 260 board feet 8 inch channel rustic siding No. 2.
- 1,250 shingles.
- 4 1x4 corner boards, each 6 feet long.
- 4 1x3 corner boards, each 6 feet long.
- 5 1x3 door and door opening, each 12 feet long.
- 2 1x4 14 feet long cornice finish.
- 4 1x4 6 feet long cornice finish.
- 1 1x3 14 feet long ridge board.
- 1 1x4 14 feet long ridge board.
- 5 1x4 16 feet long miscellaneous use.
- 1 1x2 14 feet long stops for oil-can nests.

Hardware

- 6 lbs. 8D case.
- 10 lbs. 8D common.
- 3 lbs. 16D common.
- 4 lbs. shingle nails.
- 1 pair of strap hinges.
- 6 feet of heavy wire.
- 18 feet of 1 inch mesh wire for door and front.
- 8 10x10x15 oil-cans for nests.
- 2-3 of one end cut-out.
- 4 ½x10 anchor bolts.

The Nests.—Nests for laying hens should be somewhat secluded, for fowls are less liable to acquire the egg-eating habit when the nests are in a darkened place. They should be from 10 x 12 to 12 x 14 inches in size and 8 to 10 inches high, the larger breeds requiring the larger size. A cheap and serviceable nest may be made out of a five-gallon oil can by cutting the end out, leaving about 3 inches at the bottom to keep the nest material in the nest. Such a nest can be easily cleaned either by scalding or spraying. The illustration shows top of can taken off; this makes the nest more roomy. Several of these nests may be set on a platform about 2 feet from the floor, turning the entrance of the nest toward the wall and leaving a space of 8 inches between the nest and the wall for the hens to walk along. The nest platform should be nailed to a cleat on the side of the house and braced from top of sill. Over the nests, to keep the chickens from standing on them and to help to darken them, is fitted a sloping top. This top should be built high enough so that the attendant can see into the nests from the rear. Ten nests to fifty hens should generally be provided.

Another plan for nests frequently adopted is to place them under the droppings platform high enough to permit the hens to have full use of the floor. If this plan is followed it will be necessary to raise the platform 3 feet from the floor. This is higher than desirable, especially for the heavier breeds, as they are liable to injure themselves in jumping to the floor from the roosts. However, there is little danger from this in a house without a floor or with a floor if it be covered deeply with litter as it should be. The coal oil can nests may be used under the platform, or a row of nests may be made with lumber. Whatever is used, they should be made in a way that they may be easily removed for cleaning and disinfecting.

Another plan for nests more desirable than either in mild sections where the fowls are out of doors all the year, is to put them outside the house either on the end or side of the house least exposed to rains or the hot sun. Still an-

NESTS UNDER THE DROPPING PLATFORM

The front board is hinged at the bottom and is shown open for gathering the eggs.

other plan is illustrated in a Utah colony house. In this case the nests are placed in the back wall of the house. The hens enter from the inside, while eggs are gathered from the outside.

Separate Laying House.—Where the colony system is used, as in Petaluma, Cal., a separate laying house has many advantages. It may be used in part for feed storage and feed hoppers. The space in the roosting house is all used for roosting or taken up with perches; the nests must either be on the outside wall of the house or in a separate house for that purpose. On the large Petaluma ranches no scratching houses are used, dependence being placed on the free range furnishing the necessary exercise. With heavy breeds, however, scratching sheds should also be provided.

Broody Coop.—Where an empty pen or yard is not available, a broody coop should be provided for the broody hens. This may be made from an ordinary box with a floor of slats. The slats make it cleaner and also prevent the hen sitting. Cold air circulating underneath will also help to overcome the brooding tendency. At the first symptoms of broodiness the hen should be removed to the broody coop,

A GOOD BROODY COOP
On farm of H. A. George, Petaluma.

unless wanted for hatching. This coop may be hung on the wall inside the house if there is room enough, otherwise it may be hung outdoors on the shady side of the house or in some other convenient place. If there is a vacant yard available it is a good plan to use that for breaking up the broody hens. Where large numbers of fowls are kept and broodiness becomes a considerable problem, a separate house built for that purpose may be used, such as illustrated above.

The Trapnest.—The main or essential points in a good

trapnest are simplicity, cheapness, and accuracy in operation. The Oregon Station trapnest has been in use 12 years (Utah Station Bulletin 92 and Oregon Station Circular 4). As the hen enters this nest her weight closes the door, making it impossible for her to get out or another hen to enter. The opening in the nest is made just large enough for one hen at a time to enter.

It is necessary to visit the nests several times during the day to release the hens, and there should be enough nests so that there will always be some vacant, otherwise eggs are liable to be laid on the floor. For a flock of fifty hens, 10 or 12 nests will be sufficient if they are visited often enough.

The nests may be built singly or in groups. They may be set in the wall of the house, or inside the wall. They may also be made and set up outside, separate from the house. It is sometimes an advantage to release the hens from the top instead of through the door. This can be done where there is only one tier of nests. Occasionally a hen is slow in coming to the door to be let out, and by pulling the nest out or raising the cover, the operation of releasing the hens may be more quickly performed. With the small, active breeds there is not much trouble on this score. They come quickly to the door. The heavier breeds, like the Plymouth Rock, usually take their time in coming out, and sometimes have to be pulled out. Where they can be reached from the top this trouble is overcome.

The dimensions given are for small fowls and medium-sized fowls up to about six pounds. It will be necessary to add an inch or two to the dimensions for the large breeds and increase the size or width of opening for the door.

How to Make It.—The Oregon trapnest can be made by any one who can use a saw and drive a nail. It can be cut out of a 12-inch board, 10 feet long. The material consists of: one board 1 x 12 inches x 10 feet; six screw eyes,

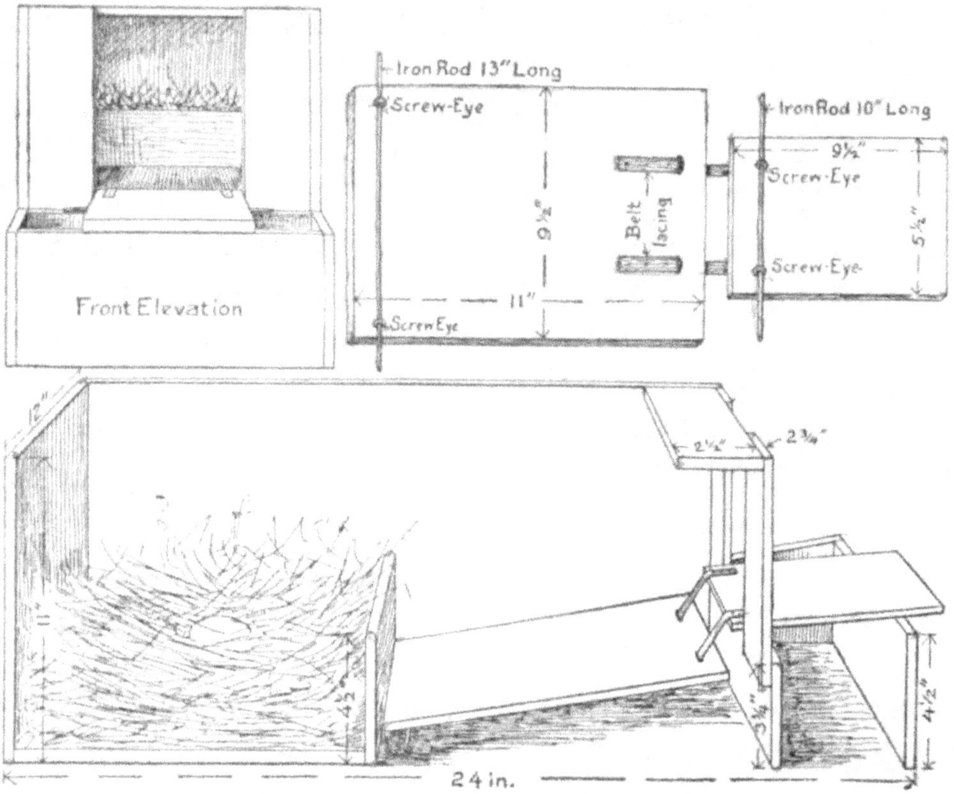

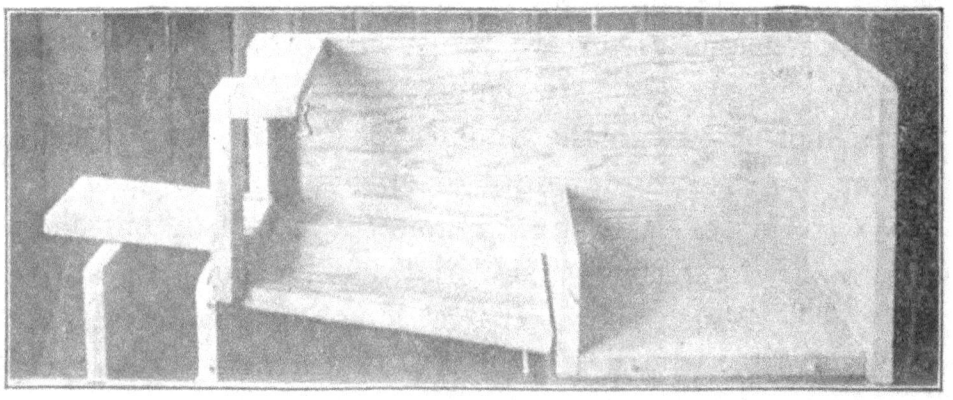

THE OREGON STATION TRAPNEST
Trapnesting is revolutionizing poultry breeding.

KIND OF HOUSE TO BUILD

No. 210 bright; two pieces of iron rod 3-16 x 12 inches, and two pieces belt lacing 9 x ½ inches.

Yards.—If the poultry houses are located near a neighbor's fields or yards, it will be necessary to yard the fowls. For other reasons, such as the keeping of more than one variety or strain of fowls, separate fenced enclosures must be maintained. Where these reasons do not exist, it is better to give the fowls free range, either in large or small flocks or in large or small houses. They will do better running together, as many as 500 in a flock, on free range than if separated into yards with fifty or 100 in each. The house may be divided into pens with partitions between each, and 50 or 100 fowls in each pen. When once accustomed to their pen they will usually go back to their own roosting places.

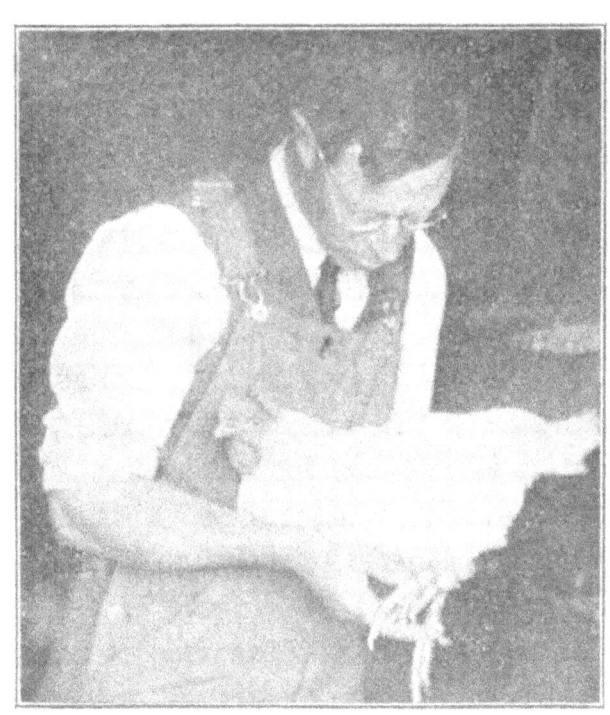

LADY MACDUFF
Being taken from the trapnest when she laid her 303d egg.

Importance of Keeping the Yards Clean.—When chickens are confined throughout the year in yards, care must be taken to keep the yards clean, otherwise there will in time be serious losses from diseases and general loss in vitality. When they are kept year after year on the same ground the yards sooner or later become contaminated with disease-producing germs, and losses through sickness and decrease in vitality will render it unprofitable to keep fowls. Dr. Salmon says: "Accumulations of excrement harbor parasites, vitiate the atmosphere, and breed contagion." It may be possible, but it is doubtful, to keep yards sufficiently clean by disinfection and other means to prevent troubles of this kind. At any rate, the expense of disinfection and cleaning would render it impracticable.

Size of Yards.—The size of yards will be governed largely by, first, amount of land available; second, nature of the soil; third, the cost of fencing; and fourth, number of separate breeds or breeding yards.

As to the first, the larger the yard the more exercise the fowls will take. Large yards, therefore, mean greater vigor in the stock. Where the soil is dry and porous with plenty of sunshine, probably double the number of fowls can be kept on the same area or yard as where the soil is heavy and wet. The larger the yard, the better for the fowls; but it is possible to make them so large that the cost of fencing will offset the advantages. In other words, the fencing becomes prohibitive when a certain limit of yard is exceeded.

The main, if not the only excuse for small, separate yards, is for keeping distinct strains or breeds separate for breeding purposes. Where as many as 500 fowls are kept and there is no object in making up small breeding pens, one large yard may be fenced in and the fowls allowed to run together in the yard. So far as there is any reliable data or experiments, the results in egg yield will be practically as good as where they are separated into small yards of 50 or 100 fowls. Again, the larger yard is more easily cultivated and cropped than small yards.

Fencing is expensive, and if the yards are very large the cost may exceed that of the houses. It requires more fencing to fence a given area in a rectangular yard than in a square yard. The estimates of yard space vary from 20 to 100, or more, square feet per fowl. For 100 hens the size of yard under favorable soil conditions, should not be less than 20 square feet per fowl with a double yard, making 40 square feet as a minimum.

Double Yards.—Where fowls are kept yarded the only practical method of keeping the yards clean or to lessen the danger of soil contamination, is to furnish at least two yards for each flock. If the long, continuous house be used and it is divided into small pens it will be better to have the yards on each side of the house, rather than two yards on one side, in order to get width enough in the yards for cultivation. The yards being shorter and wider, less fencing will be required. Where the yards are too small for horse cultivation, spading will have to be resorted to.

Portable Fence.—Portable fences may be used, such as illustrated on page 204. When the fowls are moved from one yard to another the fences are moved, so that half the amount of fence is needed, as for permanent fences. They take half the amount of wire, save the digging of post-holes and the cost of posts. In the case of continuous houses, with yards on each side, the fence is moved from one side of the house to the other, leaving the old ground open and free for cultivation. This saves in the cost of labor in cultivating. More labor, however, is required to build the portable fence, and the moving of them once or twice a year is likely to damage them somewhat, but if built of good heavy material they will last a number of years.

Portable System for the Farm.—On the general farm where 50 or 100 hens are kept, the portable fence plan may be used to advantage where it is necessary to fence in the fowls, as, for example, during seeding time and while the grain is getting a start. Part of a grain field may be used for a poultry house and flock. During part of the spring a quarter of an acre may be fenced off with a portable fence, and the flock put in a portable house, such as illustrated below. The flock would be turned loose on free range during the summer, and the following year the house and fence moved onto fresh ground. A fresh quarter acre should be given them each year for four years and in the fifth year

PORTABLE FENCE
Designed and used at Oregon Station.

they would be put back on the original quarter to follow the same rotation. The manure from the 50 fowls would keep the acre of ground in good fertility for the growth of crops, and soil contamination, with consequent diseases in the flock, would be practically eliminated. If the ground is fairly dry and the flock be not kept shut in the yard more than three months, 100 fowls could be kept on the same acre, using two colony houses. In northern sections where snow covers the ground two months or more in the winter, additional scratching room should be provided in the form of a cheap shed illustrated on page 183.

The farm flock of 50 or 100 fowls could be made the unit of larger and extensive plants. For every 100 fowls an acre of ground with two colony houses and a portable fence would be required. A system of this kind followed in combination with grain growing may be conducted with practically no cost for land. The chickens will do little if any damage to the grain crop, if the crop is pretty well grown before they are turned into it. They will eat some of the grain, but the grain will not be wasted and when the crop is harvested they will pick up the waste grain in the field. The

SHOWING HOW FENCE MAY BE CONSTRUCTED

house would then be moved out farther into the stubble field.

Fencing.—Evolution and poultry breeding have not yet produced the hen without wings. In some of the heavy meat breeds the wings are of comparatively little use. A very low fence serves to confine them. The wings of the tame duck are practically valueless to protect them from the wild animals of the forest, which was the particular use or purpose of wings in their wild state. Long disuse has lessened their power of flight and put them practically out of commission. The turkey, more than any of the domestic fowls, retains the power of flying. This is another instance of where the poultry breeder is helpless in changing the nature of the hen. In another thousand years or two the wings of fowls, through disuse, may diminish in size and strength to such an extent that poultry fences will be cut down to a height that will serve only to keep the hen from walking over them. As it is, the wing is a part of the hen which, no matter how valuable it may originally have been to her, is now positively a detriment not only to the poultryman, who is making a considerable investment in fences, but to the neighbor engaged in gardening.

The Height of Fences.—In practice, fences are usually made from 4 to 6 feet high, the lower fence for the heavier meat breeds and the higher for the light breeds. Even 6 feet is too low for some of the Leghorns or other small breeds, but fences are seldom made higher than this. By clipping the wings, or one wing of each fowl, the 6-foot fence will be high enough for the smallest or most active fowls. Where it is not desirable to clip the wing, it will be necessary to make the fence about 8 feet high for the active breeds.

Material.—Poultry fences are almost invariably made of poultry netting. It is made of galvanized wire and the size ranges about 18-gage to 20-gage, usually 19 or 20. For a durable, substantial fence, the 18-gage is recommended. The durability of the wire depends upon its being well galvanized. For adult fowls 2-inch mesh wire is used; for small chicks 1-inch or ¾-inch mesh. The posts should be set 10 feet apart, not more than 12. A 2 x 4-inch post treated with a preservative is heavy enough, though a 4 x 4 will last longer.

Shade and Fruit Trees.—Shade is very necessary for fowls in summer. This may be secured from fruit trees or other trees. Two or three fruit trees—such trees as will do best in the particular soil and location—in each yard will furnish some revenue, as well as shade. Most varieties of fruit do well in poultry yards. Prunes, apples and cherries do exceptionally well. The droppings fertilize the trees and the poultry aid materially in keeping in check certain of the fruit pests. Sour apples should be fed sparingly to fowls. Sour varieties of apples should not be planted. Where it is not desirable to plant trees, sunflower or corn may be planted early in season in part of the yard, fenced off temporarily. The sunflower is a rapid grower and furnishes excellent shade.

SUMMARY OF CHAPTERS IX AND X

1. No one style of house is essential to a good egg yield.

2. Good egg yields have been secured in long houses, and in small, portable colony houses, but the highest records have been made in the latter.

3. On one point all experiments agree, that is, the necessity of an abundant supply of fresh air.

4. Even in the cold climates of Maine, Canada, and Minnesota the cold fresh-air houses have given better results than warmly built houses.

5. Fowls require shelter more than house—shelter from winds, rains and snow, rather than from cold.

6. The open shed, or the open-front house, is the most serviceable house that has yet been invented. Without it the poultry industry would have gone to the bad before now.

7. As to how much of the front should be open will depend largely upon weather conditions. The opening may be smaller in cold climates than in warm. Additional ventilation should be given during summer.

8. Samples of air should be taken at night with the nose to determine whether the fowls are getting enough of the cheapest and best poultry food on earth—fresh air. A good nose, therefore, is part of the equipment of a poultry farm.

9. Records are not much in favor of movable curtains. It is doubtful whether they are necessary or desirable in any section.

10. Portable houses render the control or prevention of diseases much more easy.

11. After all, the house is not guilty of all the things that have been charged against it. Probably the yards should more often get the blame. A good house should not be hitched on to an unsanitary and poorly kept yard.

PRESERVATION OF POULTRY MANURE

Poultry manure has a high fertilizing value. It is especially rich in nitrogen. Unlike farm animals, fowls pass the urinary excretions in the droppings. The urine is rich in nitrogen as well as in potash, and this accounts for the high fertilizing value of the droppings.

The average fowl produces at night about thirty pounds of manure in a year. This varies somewhat as the method of feeding varies. Fowls fed a soft mash in the evening produce more manure at night than fowls that have whole grain as the last feed of the day. The night droppings, on the average, based on the value of commercial fertilizers, should be worth 15 to 20 cents per fowl; or at the rate of $30 for 100 hens during the year, counting both night and day droppings.

A large part of the value of the manure, however, will be lost unless some care is taken to preserve it. Much of the loss will be prevented if the droppings be mixed with dry loam. If stored in a shed or in barrels there should be alternate layers of loam (not sand) and manure in the proportion of about 2 inches of the former to 1 inch of the latter.

There are other methods of preserving the fertilizing constituents of the manure. One is to use gypsum. It is a pretty good plan to sprinkle the dropping board with gypsum and then mix more of it with the manure when stored. In experiments at the Maine Station it was found that "from the dung stored by itself or with sawdust, more than half of this had escaped during the summer. The lot stored with 40 pounds of plaster lost about one-third, while the lot stored with 82 pounds plaster and 15 pounds sawdust suffered no loss." The best preservation was secured with kainit and acid phosphate, both with and without sawdust. For a flock of one hundred hens a good method of preserving the manure would be to use about thirty pounds of acid phosphate or kainit to about half a bushel of sawdust. Good dry earth or muck will take the place of sawdust. Lime and wood ashes should not be mixed with the manure as they accelerate the loss of nitrogen.

CHAPTER XI

FUNDAMENTALS OF FEEDING

Feeding is one of the very important subjects in poultry husbandry. It is true that some hens will not lay many eggs, no matter how well they may be fed; that is because they have not the inherited ability to lay; in other words, they have not the proper breeding. This is discussed in the chapter on breeding. It will be seen there that food is efficient in producing eggs largely as the hen has been bred for laying, and that it is a waste to feed it to some hens. At the same time feeding must not be underrated. While it is true that some hens will not lay no matter how well they may be fed, it is equally true that some will not lay, no matter what their breeding may be, unless well fed.

The problems in feeding cannot be settled by a set of rules and regulations. That is to say, any system of feeding cannot be followed blindly under all conditions. If fowls were all alike, if climatic conditions were always the same, if foods never varied in composition, if the feeding were done with a single purpose, it might be possible to reduce the problem of feeding to one simple ration and one single way of feeding. If conditions were always the same it would be possible to say to the poultrymen in effect: Feed this ration and follow this system of feeding and you will be successful.

The successful poultryman of course will follow a system, but no system will relieve him of the necessity of doing a little thinking for himself if he will get the best value from the foods he feeds. His success in securing a good egg yield and, therefore, a good profit will depend very largely upon his knowledge of foods and the skill he exercises in feeding.

That the fowls of to-day lay considerably more eggs than their wild progenitors did is due in part to better feeding and a more abundant supply of food. But the hens are not laying on the average half what they should. To secure the maximum egg yield the poultryman must give earnest attention to the feed bucket and to methods of feeding. High success in securing eggs can only come where the subject of poultry feeds and feeding is given earnest study.

A Knowledge of the Composition of Foods will enable the poultryman to gain a clearer conception of their values. The advance in poultry feeding in recent years has been due in part to a better knowledge of the composition of foods. While our knowledge of poultry foods and feeding may never reach a point where we can say that certain foods or rations will produce certain results, yet a great deal of valuable information is available as a result of experimental feeding at the stations and of chemical analysis of poultry foods. In addition we have the experience of practical poultry-keepers, which constitutes a fund of valuable information to draw upon. But poultry feeding has not yet been reduced to a so-called scientific basis. While this is true, the student of poultry feeding will be agreeably surprised to find much data of such a character as to well repay diligent study and research. The manufacture of eggs—for egg production is really a manufacturing process, the hen being the factory—requires a careful study of the raw materials as well as of the finished product, and the working of the factory itself. If the poultryman wishes to achieve the highest measure of success, it is imperative that he avail himself of the information that is available as a result of costly experience and experiment.

Limitations of Feeding.—Most poultry-keepers do not realize the importance of good feeding; others place the whole responsibility upon the food and feeding. Before telling what food will do, let us first tell what it will not do; let us understand some of its limitations

First.—Good food and good feeding will not make some hens lay; they are not bred to lay. At the Oregon Station one hen laid 259 eggs in one year; another, fed on the same food, laid six eggs. In another case one hen laid 268 and a flock mate on the same ration laid three. Many other similar instances might be given. This is referred to in detail under the chapter on breeding.

Second.—Good feeding will avail little unless the fowls have good housing or care, or, in other words, favorable environment.

With good fowls and good housing, what will good feeding do in the production or manufacture of eggs?

Food Affects the Quality of Eggs.—The hen is very particular about what she puts into the egg, so particular that probably no food could be fed that would render the eggs totally unfit for consumption. At the same time it has been demonstrated by experiment that food affects the quality of the egg, and that to produce eggs of the highest quality attention must be paid to the quality of the food.

Flavor of Eggs.—Heavy feeding of onions, for example, will give a distinct flavor to the eggs and make them almost unpalatable. Hens eating large quantities of beef scrap will lay eggs of strong flavor. These facts the writer personally demonstrated by experiment. No doubt other foods will also give a flavor to the eggs, desirable or undesirable. It is said that a diet of fish will give a fishy taste to the eggs.

It is not necessary, however, to discard these foods on this account, for when fed in normal quantities they will not give a perceptible flavor to the eggs. Only when the hens have been starved on green food or animal food, and then given all they will eat of either for a few days, will any flavor from onions or animal food be noticed in the egg. But this shows that the hen puts into the egg what she finds in the food, even the flavor of the foods. It is therefore important that good wholesome food be fed at all times.

Feeding Color Into the Egg.—It is possible for the skillful feeder to flavor the eggs; it sometimes happens from unskillful feeding, as indicated above. It is possible also to "paint" them. The variation in the shade of yellow in the yolk is due to a difference in the food. The coloring of the egg shell is beyond the feeder's art, but food affects the color of the yolk as we have demonstrated. A pen of fowls fed dried alfalfa leaves produced eggs of good yolk color. A similar pen fed sugar beets instead of alfalfa leaves laid eggs very pale in color. In an experiment at the Oregon Station kale "painted" the yolks a good color of yellow. Experiments at other stations have shown that the feeding of yellow corn will color the yolk. (West Virginia Bulletin 88.) When eggs are pale in the yolk it is a sure indication that the hens are not getting green food enough. Clover, vetch, rape, grass, or other green food, and doubtless certain grain foods, will color the yolk. A yolk too highly colored is not desirable, and it is possible for the hens to eat so much of certain foods as to color it too highly. Where the ration is right this should not occur. Food, therefore, affects the quality of the eggs.

It has been further demonstrated that it is possible to color both the yolk and the white of the egg by the feeding of certain aniline dyes. Rhodamine Red dye fed at the rate of 100 grams daily will, in a few days, color the white a pink color, while Soudan III dye will in about two weeks of feeding color the yolk a dark red. An egg laid two days after feeding this dye to the hen will show the outer rim or layer of the yolk colored. An egg laid at the end of two weeks of feeding will show each layer of yolk distinctly colored.

So far the experiments referred only to color and flavor. Both color and flavor in the egg are points that have a market value. Eggs either too pale or too highly colored in the yolk will be objected to by consumers who pay a fancy price and expect a fancy article. So, too, the flavor must be unobjectionable if fancy prices are to be received.

These experiments might indicate that it is possible, by feeding certain foods, to change the chemical composition of eggs or feed into them certain things that will improve their nutritive value. So far, however, this is only a possibility. Little investigation has been done and what has been done seems to show contradictory results.

Investigations by Cross at Cornell showed that "in feeding a ration high in fat or a ration high in protein there is no material change in the amount of fat and protein in the egg." There is need, however, of further investigation and it would seem that the matter is of practical importance enough to warrant it.

Food Affects the Yield of Eggs.—Other conditions being right, good feeding makes the hen productive, and the productive hen is the healthy hen. In a pen of four fowls at the Utah Station 804 eggs were laid in one year. Another pen of four, sisters to the others, fed a different ration, laid 532 eggs. The difference in the ration made the difference in the egg yield. In another test one pen laid 574 eggs in a year, and a similar pen on a different ration laid 404.

In a West Virginia experiment fowls fed a nitrogenous ration laid 7,555 eggs, while other fowls fed a carbonaceous ration laid 3,431 eggs. (West Virginia Bulletin 60.)

FUNDAMENTALS OF FEEDING

Food Affects the Size of Eggs.—Food and feeding influence the size of eggs. Do not always blame the hens or the breed for small eggs. An experiment has shown that the size of egg is influenced by factors under the control of the poultryman.

The size of egg, of course is influenced by other factors. The size varies to some extent as the vigor of the fowl does, and vigor is very largely dependent upon the food and method of feeding. This fact was brought out in an experiment by the writer at the Utah Station. Fifty Leghorn pullets were divided into four lots, as follows:

Pen 2, 10 fowls. In a continuous house, closed front, slightly artificially heated.
Pen 14, 10 fowls. In a continuous house, closed front.
Pen 26, 10 fowls. In a continuous house, open front.
Colony house, 18 fowls. On free range.

The average weight of eggs for the six months beginning December 1, was as follows:

Colony house .... 25.3 ounces per dozen
Pen 2 .... 23.4   "   "   "
Pen 14 .... 23.5   "   "   "
Pen 26 .... 22.5   "   "   "

Eleven eggs from the colony house, it is seen, weighed as much as 12 from the other pen. The increased size of the eggs from the colony house flock was due to one or two factors, or to both, namely, to greater exercise and natural foods secured on the range. It was not a question of fresh air or type of house, because in the open-front house the eggs were no larger than those from the closed-front house. That there is a relation between the size of egg and vigor of the fowl is evident from the fact that the fowls in the colony house and on free range weighed heavier than those in the other houses at the end of experiment, though their weights were equal at the start. The size of egg is undoubtedly influenced by the physical condition or vigor of the fowl. The food affects the vigor of the fowl and therefore affects the size of egg. In more recent experiments at the New Jersey Station, rations deficient in protein produced undersized eggs. (New Jersey Bulletin 265.) Other recent experiments at West Virginia indicate that scanty feeding produces undersized eggs (West Virginia Bulletin 145).

Food Affects the Profits.—A proper study of foods and feeding must include prices as well as composition. A ration, although it may give good results in egg yield, may not be profitable because it is made up of too high-priced foods. There is no patent on egg-producing foods. It is not necessary to use any certain kind or brand of foods. It is not necessary to pay more for the chickens' food than for the food for the family table. There are rations that are impracticable because they are too high-priced.

Different Elements of Food.—If we look upon the hen as a factory for the production of eggs and the eggs as the finished product, the food will be the raw material. If we had never seen a hen eating wheat we should hardly suspect that eggs were made out of wheat. Eggs and wheat do not look much alike, and yet when the chemist analyzes them he finds that they are pretty much alike in composition. The farmer manufactures wheat from the soil, with the assistance of the heat from the sun and the rain from the clouds. The crop it produces he separates into straw, chaff and grain. The chemist takes the grain and separates that into water, protein, carbohydrates, fat and ash. The poultryman feeds wheat to the hen and the hen produces eggs. The chemist analyzes or separates these eggs as he did the wheat and he finds that they contain the same elements as he found in the wheat, namely water, protein, carbohydrates, fat and ash, with the difference that the carbohydrates have been converted into fat. The main difference between a bushel of wheat and a bushel of eggs is that the eggs are more palatable and more nutritious. They are also more valuable in the market.

A study, therefore, of the composition of the finished product gives us a clue as to what the raw material should be.

Composition of Eggs.—Without the shells a dozen eggs weighing 1½ pounds, contained 13.57 ounces water, 2.32 ounces protein, 2.26 ounces fat and 0.22 ounces ash. A pound of eggs is worth from 10 to 30 cents, depending upon the season and markets; a pound of wheat runs from 1 to 2 cents. When wheat is given to the hen it is converted by a delicate process of manufacture into a form of food so valuable that it is worth many times as much as it was in the grain sack. More than that, the hen is thrifty; for every pound of wheat she puts into eggs she puts a pound of water, as will be seen later; and she gets a good price for the water. In selling eggs at 40 cents a dozen the poultryman is getting 25 cents a pound for the water in them. It is more than the dishonest dairyman gets for the water he puts into his milk.

An average egg weighs two ounces: 10.81% of it is shell, 32.47% of it is yolk, and 56.42% of it is white.

The yolk is composed of about 50% water, 15.5% protein, 33.4% fat and about 1% mineral matter.

The white is composed of about 85% water, 12.1% protein, 0.23% fat and 0.34% mineral matter.

Relation of Food Eaten to Eggs Laid.—There is a close relationship between the character of the raw material or food and the finished product. The skill of the poultryman comes in in properly adjusting the ration to meet the requirements of heavy production. The hen does not adjust the composition of the egg to the food that may be fed. If the right elements are not present in the food she refuses to make eggs. The composition of the egg does not vary to any extent.

The egg contains one-quarter ounce of protein. If the hen be fed on wheat and nothing else she may eat four ounces per day. Of that she will need about three ounces to supply bodily needs. This leaves one ounce to make eggs with. In an ounce of wheat there is about one-tenth of an ounce of protein. Now, supposing the protein is all digested, which is not the case, she will not get enough protein to make half an egg a day. But an egg every two or three days would not be so bad at certain seasons. The egg, however, contains other things. It contains also about one-quarter ounce of mineral matter, chiefly lime for shell. An ounce of wheat contains less than one-tenth as much mineral matter as one egg of two ounces contains. The egg also contains fat. It contains less than one-quarter ounce of fat, but the wheat would contain three-quarters of an ounce of fat formers.

What would be the result if the hen were fed on wheat alone? She would get enough protein to make an egg about every three days; enough lime to make an egg every 12 days and enough carbohydrates and fat to make three eggs a day. What will the hen do in such a quandary? She could put more fat into the egg to make up for lack of protein. She could make a counterfeit article, but she will not. Unless she has the right materials to make it with, she will not make the egg. What would probably happen would be that she would lay an egg every three or four days, every two out of three soft shells, and the surplus fat and carbohydrates would be wasted or put on the hen in the shape of surplus fat. This is assuming that the hen would continue to consume four ounces of wheat a day and maintain health. In practice, however, the result would be different. She would not long continue to eat four ounces of wheat and nothing else. There would soon be a loss of appetite and health.

It is poor economy to feed wheat alone. The same thing is true of corn and all the cereals. None of them are "balanced" for egg production.

A Balanced Ration.—This raises the question of what is a balanced ration? A balanced ration is one containing the right kind of nutrients in right proportions for the purpose for which it is fed.

We must know the composition of foods before we can figure up a balanced ration. It may not be necessary in practice for the poultryman to figure up balanced rations for his flock. His experience or the experience of others, or the results of tests at experiment stations, are a pretty safe guide for the poultryman; but in order that he may intelligently plan improvements in rations, and adjust his feeding to the available food supply, he should understand something of the composition of ordinary poultry foods.

A BALANCED RATION

Wheat, oats, bran, and beef scrap in the above proportions make up a balanced ration for laying hens. In addition, green food, grit, and oyster shell must be fed.

What Use Does a Hen Make of the Food She Eats?—In other words, what is the purpose of feeding? The first use she makes of the food is to supply the needs of her body. The maintenance of her body is her first concern. The body of the hen, like that of other animals, needs constant rebuilding. There is constant wearing or breaking down of tissues, and the food rebuilds the body or repairs its wastes. The work of the poultryman, therefore, does not end with the making of the hen, with the hatching and rearing of the pullet; he must maintain her, and the skill of the feeder shows itself in so compounding rations and so feeding them that the health and vitality of the hen may be maintained. That is the first consideration of good feeding—the maintenance needs of the hen, the maintenance of health and vigor.

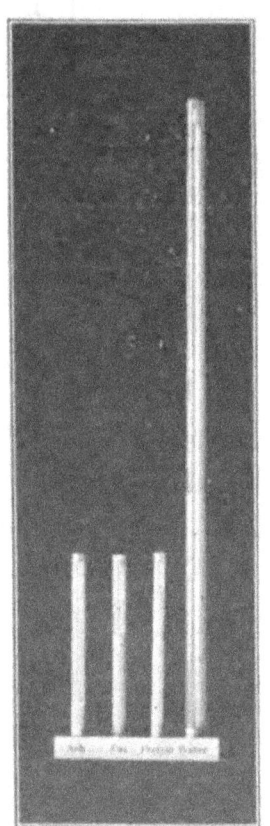

The relative amounts of ash, fat, protein and water in eggs.

In feeding laying fowls, the second use to which food is put by the hen is to make eggs. After the body's needs have been supplied, if there is any food left, the hen will use it for the making of eggs. Eggs are made from surplus food. After she has eaten enough to supply bodily needs she turns attention to the egg basket. It is poor economy, therefore, if the purpose is egg production, to feed just enough to maintain the hen. More must be fed or our efforts will be wasted.

If the purpose is meat production and a fattening or fleshening ration is being fed, the purpose will be defeated if only enough is fed to maintain the fowl. The profit in feeding in both cases comes from the food consumed above that necessary for maintenance.

On the other hand, heavy feeding does not necessarily

mean a heavy yield of eggs. In an experiment by the writer two pens of fowls consumed an average of 75.6 pounds food, not counting the green food, and laid an average of 167 eggs per fowl. With the same amount of food two other pens averaged 117 eggs each. The nutritive ratio was practically the same in each case. While the heavy layer must consume plenty of food, the manner of feeding and the kind of food must be taken into account. In other words, the efficiency of feeding rests largely on the kinds of food fed and the skill with which the feeding is done.

FEED REQUIREMENTS OF CHICKENS PER DAY FOR EACH 100 POUNDS OF LIVE WEIGHT (AFTER WHEELER)

	Digestible nutrients (pounds)				
	Protein	Fat	Carbo-hydrates	Ash	Total dry matter
Growing chicks:					
First 2 weeks	2.00	0.40	7.20	0.50	10.1
2 to 4 weeks.....	2.20	0.50	6.20	0.70	9.6
4 to 6 weeks.....	2.00	0.40	5.60	0.60	8.6
6 to 8 weeks.....	1.60	0.40	4.90	0.50	7.4
8 to 10 weeks....	1.20	0.30	4.40	0.50	6.4
10 to 12 weeks...	1.00	0.30	3.70	0.40	5.4
Adults (maintenance only):					
Capon, 9 to 12 pounds	0.30	0.20	1.74	0.06	2.3
Hen, 5 to 7 pounds	0.40	0.20	2.00	0.10	2.7
Hen, 3 to 5 pounds	0.50	0.30	2.95	0.15	3.9
Egg production:					
Hen, 5 to 8 pounds	0.65	0.20	2.25	0.20	3.3
Hen, 3 to 5 pounds	1.00	0.35	3.75	0.30	5.4

Food Requirements.—The food requirements vary with the age and size of the fowls. The younger the chick the more food is required per pound weight of chick. The larger the laying hen, less food is required per pound weight of hen. It has been shown in experiments by Wheeler that 100 pounds of chicks under two weeks of age required 10.1 pounds of food (digestible nutrients) per day; from four to six weeks the requirement was 8.6 pounds; at 10 to 12 weeks the requirement was 5.4, so that according to the weight of the chick, or for every 100 pounds of chicks regardless of number, nearly double the amount of food is required during the first two weeks of their age as is required from the 10th to the 12th week. It is also shown that the small chick requires double the amount of food that the laying hen needs, per pound weight. For a hen not laying, the difference is still greater. More food, of course, is eaten per chick as it grows older, but less is eaten per pound weight of chick.

Natural and Artificial Feeding.—The business of poultry keeping is more or less artificial, even the feeding of the fowls. Artificial methods, however, can be followed successfully just so far. The lessons of feeding will be more easily learned if account be taken of the manner in which fowls secure their food under natural conditions. Where they have their liberty to range over fields they pick up weed seeds and waste grain, nibble at the grass and grass roots, chase flies and grasshoppers, hunt for bugs and worms, and finish off with grit for dessert. Under such conditions the hen balances her own ration, maintains her health and vigor and produces eggs abundantly, if the supply of these foods is large enough so that she can secure her meals regularly each day. The exercise secured in hunting for the food enables her to better digest and assimilate her food and maintain her in good health and vigor. But under natural conditions the daily food supply is uncertain, and here is indicated the advantage of artificial or systematic feeding, or the necessity of cooperation between the farmer and the fowl, if the highest production is to be secured.

The Purpose of Feeding is not merely to maintain the fowl in health and vigor; she can take care of that herself if given her liberty; the purpose of feeding is to secure higher production, and that is possible only where the food supply is sufficient and regular for the needs of the hen. Account must be taken of the nature of the hen. She must be fed artificially, but artificial foods or nutrients must not be substituted for the foods obtained naturally. Neither may a life of ease be substituted for her natural life of activity. She is a creature of great nervous activity and the poultryman must take account of that also and in his feeding make sure that the activity or exercise is provided. Nature calls for food of certain kinds and for activity or exercise that will make the food efficient in production. We cannot improve on the kinds of foods, nor do away with activity. But the intense production called for in the modern improved egg-producing hen calls for systems of feeding that will furnish unfailingly a full supply of all the food nutrients demanded by the fowl.

Composition of Foods.—This does not mean that the feeder must limit himself to weed seeds and bugs and grasshoppers. Wheat and corn are made up of the same ingredients as wild weed seeds, namely, protein, fat, carbohydrates; so the modern meat scraps contain the same elements as grasshoppers and worms. The difference is that we furnish the vegetable protein, carbohydrates and fat, in the form of wheat and corn instead of weed seeds, and the animal protein and fat in the form of meat scraps rather than in the form of bugs and insects.

The Mineral Matter called ash, which is that part of the food that remains after burning, is found in varied amounts in all foods. The hen is a concentrator; she takes the mineral in the food, concentrates it into egg shells and mixes a little in the contents. All grain foods contain insufficient lime with which to make egg shells and the laying hen must eat grit, oyster shells, or other things, to supply the deficiency. Where high egg production is called for, the mineral matter is a most important part of the food. Its importance should be more fully emphasized.

The Oregon Station hen that laid 42 pounds of eggs in 12 months used in the manufacture of shells practically 3½ pounds of lime. In addition there was a small quantity of mineral matter in the egg contents. The grain foods she ate contained about two pounds of ash. More than half the mineral matter, therefore, was secured from other sources than the grain foods.

Mineral nutrients are also demanded by the fowl for building up or repairing the bones or skeleton of the body. The flesh and internal organs also contain certain compounds of ash. The importance of ash in feeding has been brought out in feeding experiments with hogs. Corn alone, which is low in mineral matter, produced small gain in weight and developed an undersized, fine-boned, over-fat animal "characterized by proportionately small kidneys, lungs, heart, liver and muscles, and by a high percentage of fat." A German physiologist proved that animals will live longer with no food at all than with food containing no mineral matter. As to the effect of insufficient calcium (lime) Sherman quotes the following: "Voit kept a pigeon for a year on food poor in calcium without observing any effects attributable to the diet until the bird was killed and dissected, when it appeared that, although the bones concerned in locomotion were still sound, there was a marked wasting of lime salts from other bones, such as the skull and sternum, which in places were even perforated. The injurious effects of an insufficient intake of lime is, of course, more noticeable with growing than with full-grown animals."

As the egg-producing capacity of fowls is improved there is increased demand for the mineral elements, and the successful poultryman will see that there is no deficiency in this respect in the ration. The mineral matter in the body of the fowl is largely phosphate of lime, while the egg shell is almost entirely carbonate of lime.

Ground bone is the most available form in which to furnish the mineral matter for body growth. Rapidly growing young chickens require much mineral matter in the form of lime phosphates which are found in bone. It is different in the case of the laying hen. The shell of the egg is almost all carbonate of lime and this is found in its most available form in oyster shells.

But little is known of the effects of the specific mineral elements, phosphorus, iron and sulphur, on production or growth. It will be a distinct advance in the practice of feeding when more definite knowledge has been gained of the part played in the economy of feeding by these different mineral compounds of ash.

Protein is the most valuable part of the food, because, though it is found in all poultry foods, it is not found in the cheaper foods in sufficient amount for the needs of the fowl, especially the laying fowl. Foods containing a high percentage of protein are usually the most expensive. Protein makes the lean meat and the muscle and a large percentage of the contents of the egg. The white of the egg, lean meat, gluten of the flour, and milk casein are practically all protein.

The value of the food must be determined largely by the amount of protein which it contains, and high prices should not be paid for food of any kind unless it has a guaranteed analysis of high protein content. Generally speaking, foods

226 POULTRY BREEDING AND MANAGEMENT

are cheap or dear in proportion as they contain a high or low percentage of protein.

Carbohydrates and Fats furnish the fat of the body and of the egg. From them are derived the heat necessary to keep up the temperature of the body. They are burned in the body to furnish the heat and also the energy. It requires energy to digest food; it requires energy to walk and to fly and to scratch, just as it requires steam to drive

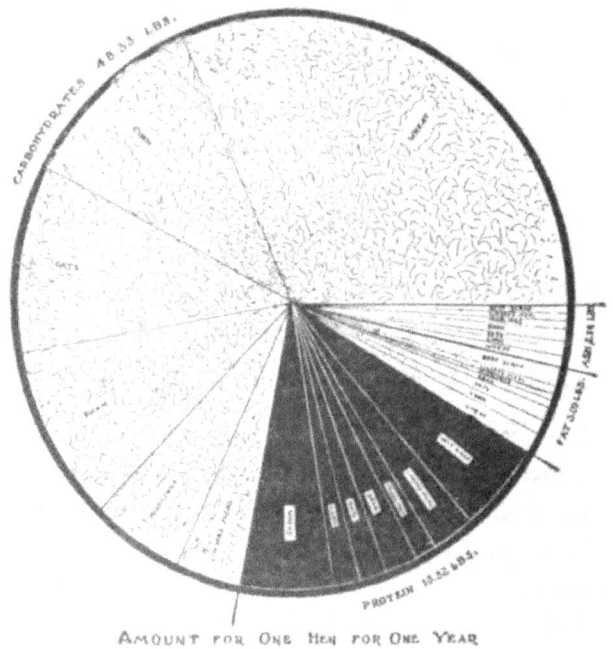

BALANCED RATION FOR ONE HEN FOR A YEAR
Showing the amount and sources of the different chemical constituents.

the steam engine; and a considerable amount of food in the form of carbohydrates and fat is used to produce this energy. Most poultry foods contain a larger percentage of carbohydrates and fat for egg production than is necessary, while there is usually a deficiency of protein. To what extent fat in the food influences the egg yield is not definitely known. In experiments by the writer rations containing a liberal amount of fat gave a better yield than others of little fat. The experiments have shown that fowls eat

PERCENTAGE COMPOSITION OF FOODS

Grain	Water	Ash	Protein	Fiber	Carbohydrates N-Free Extract	Fat
Wheat	10.5	1.8	11.9	1.8	71.9	2.1
Corn	10.9	1.5	10.5	2.1	69.6	5.4
Kaffir Corn	9.3	1.5	9.9	1.4	74.9	3.0
Oats	11	3	11.8	9.5	59.7	5
Peas	10.5	2.6	20.2	14.4	51.1	1.2
Barley	10.9	2.4	12.4	2.7	69.8	1.8
Rye	11.6	1.9	10.6	1.7	72.5	1.7
Buckwheat	12.6	2	10	8.7	64.5	2.2
Cow peas	14.8	3.2	20.8	4.1	55.7	1.4
Sunflower	8.6	2.6	16.3	29.9	21.4	21.2
Millet	14	3.3	11.8	9.5	57.4	4
Sorghum	12.8	2.1	9.1	2.6	69.8	3.6
Flaxseed	9.2	4.3	22.6	7.1	23.2	33.7
Wheat bran	11.67	5.18	14.05	8.16	57.34	3.6
Wheat middlings	11.73	2.85	15.22	4.88	60.85	4.47
Wheat shorts	11.8	4.6	14.9	7.4	56.8	4.5
Linseed meal (N. P.)	9.9	5.6	35.9	8.8	36.8	3
Gluten meal	8.1	1	28.3	1.1	50.8	10.7
Cottonseed meal	8.2	7.2	42.3	5.6	23.6	13.1
Soy bean meal	10.8	4.5	36.7	4.5	27.3	16.2
Brewers' dried grain	8	3.4	24.1	13	44.8	6.7
Green Foods						
Alfalfa	71.8	2.7	4.8	7.4	12.3	1
Clover (Red)	70.8	2.1	4.4	8.1	13.5	1.1
Kale	88.2	1.82	2.57	1.47	5.32	.61
Cabbage	90.5	1.4	2.4	1.5	3.9	0.4
Vetch	69.18	2.71	3.76	9.64	14.22	.49
Mangel Wurzel	91.2	1	1.4	0.8	5.4	0.2
Turnip	90.5	0.8	1.1	1.2	6.2	0.2
Sugar beet	86.5	0.9	1.8	0.9	9.8	0.1
Dried beet pulp	6.4	3.3	10.8	19.8	58.4	1.3
Carrot	88.6	1	1.1	1.3	7.6	0.4
Potato	78.9	1	2.1	0.6	17.3	0.1
Artichoke	79.5	1	2.6	0.8	15.9	0.2
Animal Food						
Skim milk	90.6	0.7	3.3	...	5.3	0.1
Buttermilk	90.3	0.7	4	...	4.5	0.5
Whey	93.8	0.4	0.6	...	5.1	0.1
Cottage cheese	72	1.8	20.9	...	4.3	1
Milk Albumen	18	3	43	?	?	1.5
Beef scrap	10.7	4.1	71.2	...	0.3	13.7
Cut bone	34.2	22.8	20.6	...	1.9	20.5
Dried blood	8.5	4.7	84.4	...	...	2.5
Dried fish	10.8	29.2	48.4	...	...	11.6

more food during the cold weather than during the warm. This is because it requires more food to keep up the heat of the body, and for heat-producing purposes cheap fat foods serve the purpose as well as expensive protein foods.

Nutritive Ratio.—The hardest problem, therefore, in poultry feeding is to compound suitable rations containing the necessary protein in its most available form and at reasonable cost for heavy production. The nutritive ratio is the ratio of digestible protein to digestible fat and heat-producing foods. For egg production a narrow nutritive ratio should be fed. A ratio of one of protein to four or five of carbohydrates and fat is a narrow ratio and will give good results in egg production. In figuring the ratio the fat is multiplied by $2\frac{1}{4}$ as it is estimated that one pound of fat is equal to $2\frac{1}{4}$ pounds of carbohydrates.

It should be understood, however, that the nutritive ratio in itself does not necessarily indicate the true value of the ration. Palatability and other factors have to be considered. At the Utah Station two rations having the same nutritive ratio were fed to two different pens of fowls for a year. One of them gave a yield of 201 eggs per fowl; the other 133. There was a difference in the kind of the food, but not in the nutritive ratio. Two other pens having rations of similar nutritive ratio gave yields of 101 and 143 eggs respectively. At the West Virginia Station laying hens fed a narrow ratio, or nitrogenous ration, produced 17,459 eggs, while the pens with a wide, or carbonaceous ration, laid 9,708 eggs. During the experiment the former fowls gained in live weight 1 pound 4 ounces each, while the latter gained only about one-tenth of a pound each. Other experiments have shown the superiority of the narrow ratio, or the ration rich in protein or nitrogen.

The proper nutritive ratio, however, does not guarantee a good egg yield. Regard must be had to the kind of foods fed, and the feeder must be guided by the results of feeding tests that indicate the feeding value of different foods.

It has been shown, for example, that there is a difference in the protein. Fowls require a certain amount of protein in the ration, but to be effective in egg yield part of that protein must come from animal sources. It is protein just the same, but why there should be this difference in feeding value between animal and vegetable protein is not yet known.

Experiments by Wheeler showed that an animal food ration for laying hens was superior to others in which all the organic matter was derived from vegetable sources, and for growing ducklings very much superior. In the case of growing chicks where bone ash was fed in the place of animal food the results were equally satisfactory. (Geneva Bulletin 171.)

In New Jersey experiments (Bulletin 265) it was found that: "The addition of animal protein in the form of meat scrap materially increases the efficiency of a ration relatively high in vegetable protein, both for egg production and for flesh growth," and that: "Phosphoric acid from an organic source (animal bone) is much more efficient than phosphoric acid from an inorganic scource."

ANALYSES OF FOWLS AND EGG*

(The analyses of the fowls include the feathers, bones, blood, etc.)

	Water	Ash	Protein	Carbohydrates	Fat
Hen	55.8	3.8	21.6	..	17
Pullet	55.4	3.4	21.2	..	18
Capon	41.6	3.7	19.4	..	33.9
Fresh egg	65.7	12.2	11.4	..	8.9

*Prof. W. P. Wheeler, Geneva (N. Y.) Station.

Computing the Ratio.—The nutritive ratio may be computed as follows: Suppose the ration is 10 pounds wheat, 3 pounds oats, 2 pounds bran and 1 pound beef scrap. By referring to the table of composition of feeds, page 227, it will be found that wheat contains 11.9% protein; so that in 10 pounds wheat there are 1.19 pounds protein; it contains 73.7% carbohydrates, and in 10 pounds there are 7.37 pounds carbohydrates. The percentage of fat is 2.1, or 0.21 pound fat in 10 pounds wheat. Figuring the other foods in the same way, we get the results shown in the following table:

	10 lbs. wheat lbs.	3 lbs. oats lbs.	3 lbs. bran lbs.	1 lb. beef scrap lbs.	Total lbs.
Protein	1.19	0.35	0.29	0.66	2.49
Carbohydrates	7.37	2.08	1.31	..	10.76
Fat	0.21	0.15	0.07	0.14	0.57

To get the nutritive ratio, multiply the total fat by 2¼ (0.57×2¼=1.28). Add this to the carbohydrates (10.76+1.28=12.04). Divide this by the total protein (12.04÷2.49) and we get the nutritive ratio of 1:4.8. In other words, this ration contains one pound of protein to 4.8 pounds carbohydrates and fat. This is not given as a good ration, but simply to show how the nutritive ratio is computed. In point of fact, this method of computation is not correct because it is figured on the total nutrients, not on the amount actually digestible.

Digestibility of Poultry Foods.—In the above computation it is seen, for example, that there are 1.19 pounds of protein in 10 pounds wheat, but according to Henry's compilation of digestion coefficients for livestock, there is only 0.88 pound digestible protein in 10 pounds wheat. The amounts digested are shown in the following table, using the standard coefficients for livestock:

	10 lbs. wheat	3 lbs. oats	2 lbs. bran	1 lb. beef scrap	Total
	lbs.	lbs.	lbs.	lbs.	lbs.
Protein	0.88	0.26	0.24	0.61	1.99
Carbohydrates	6.75	1.47	0.84	..	9.06
Fat	0.15	0.13	0.05	0.14	0.47

Ratio—1:5.0

Foundation of Scientific Feeding.—The composition of foods affords a means of estimating fairly well the value of the food. Foods are usually valuable in proportion as they contain a high or a low percentage of protein. For instance, a pound of protein may be worth so much, whether it be found in corn or wheat bran. A hundred pounds of corn containing 10.5 pounds protein is not worth as much as 100 pounds beef scrap containing 60 pounds protein. That is the fundamental lesson that the chemical analysis of foods teaches.

Chemistry gave to the world only some fifty years ago a feeding standard based upon the chemical composition of foods. Previous to that time, as Henry says, "the farmer gave his ox hay and corn without the least conception of what there was in this provender that nourished animals." The discovery of the vital differences in the amount of nutrients in different foods was the foundation of scientific feeding.

Percentage Digested.—But that is not all. It was found that not only did the foods vary in composition, or total nutrients, but a few years later a German scientist formulated a new standard based, not on total amount of nutrients—protein, carbohydrates and fat—but on the amount or percentage of these nutrients digested by the animal. For example, there are 3.8 pounds crude protein in corn stover, but only 1.4 pounds of that is digestible, or 36%, the rest of the protein is wasted.

58 per cent of the protein of clover is digested
76 per cent of the protein of corn is digested
77 per cent of the protein of oats is digested
89 per cent of the protein of linseed meal is digested

So do the carbohydrates and fat vary in digestibility in different foods.

The percentages of these nutrients digested by animals have been determined for practically all animal foods, and tables of digestion coefficients for livestock have been made and published. Unfortunately, the same information is not available for poultry feeds. It has been assumed that the digestibility of feeds will not be the same with poultry as with livestock; that poultry may or may not digest the food better than livestock; and that before the digestibility of poultry foods may be known digestion experiments must be made with poultry. Some work has already been done with fowls, but hardly enough to definitely establish feeding standards. So far the results indicate that the digestibility of certain foods does not vary much whether fed to fowls or to farm animals.

In the above table the figures for livestock were used in computing the nutritive ratio.

Digestion Coefficients.—This is the term used in speaking of the percentage of foods that is digestible. The digestion experiments that have been made with poultry have been mainly those by Bartlett of the Maine Station (Bulletin 184), Brown of the Bureau of Animal Industry (Bulletin 156), and Fields and Ford of the Oklahoma Station (Bulletin 46). A table of the digestion coefficients, giving the average results of all these analyses, has been compiled by Bartlett and published in the Maine Station bulletin 184. This includes the results of work of several European investigators.

AVERAGE DIGESTION COEFFICIENTS OBTAINED WITH POULTRY TO DATE

	Number of Experiments	Organic Matter	Crude Protein	Nitrogen Free Extract	Ether Extract
Bran, wheat	3	46.70	71.70	46	37
Beef scrap	2	80.20	92.60	..	95
Beef (lean meat)	2	87.65	90.20	..	86.30
Barley	3	77.17	77.32	85.09	67.86
Buckwheat	2	69.38	59.40	86.99	89.22
Corp, whole	16	86.87	81.55	91.32	88.11
Corn, cracked	2	83.30	72.20	88.10	87.60
Corn, meal	2	83.10	74.60	86	87.60
Clover	3	27.70	70.60	14.30	35.50
India wheat	3	72.70	75	83.40	83.80
Millet	2	..	62.40	98.39	85.71
Oats	13	62.69	71.31	90.10	87.89
Peas	3	77.07	87	84.80	80.01
Wheat	10	82.26	75.05	87.04	53
Rye	2	79.20	66.90	86.70	22.60
Potato	6	78.33	46.94	84.46	..

These results should be taken as more or less tentative until further work has been done and the final results based on the averages of a great many analyses.

It is noted with interest that this compilation gives a higher coefficient for corn than for wheat. If this finding should prove to be final it would mean that the value of the protein in the corn was about 8% greater than that of wheat; the carbohydrates about 5% greater, and the fat or ether extract about 66% greater. These results are not presented as final, but rather as a record to date of progress in a very important line of research. On the whole, the work indicates that the digestibility of foods may not vary a great deal whether fed to poultry or to livestock.

Palatability.—It has been pertinently said that it is possible to make a mixture of wet leather and a petroleum jelly that would give the same result as meat by the ordinary food analysis. Palatability comes in here. Leather and petroleum jelly would scarcely be as palatable as meat, nor would it be expected that the one would give as good an egg yield as the other.

Again, while a high digestibility of food is important, yet digestibility is not a certain measure of the value of the food. Sherman in "Chemistry of Food and Nutrition," says: "Foods similar in chemical composition and equally well digested, may or may not be of equal nutritive value," and again: "The coefficient of digestibility is but little influenced by the palatability of the food."

Summing Up.—We have here three factors—there may be others—that must be taken into account in arriving at the true value of a food; namely, composition, digestibility and palatability. While any of these factors, standing alone, may not mean much to the feeder, no one of them must be disregarded. In proportion as his knowledge covers all three factors he will be able to feed intelligently; but after all the knowledge that comes from practical feeding experiments is all important and necessary to a complete knowledge of the value of any particular ration.

Digestive Organs.—The organs of the fowl concerned in digestion of food are shown on p. 235. This photograph shows the various organs beginning with the mandibles or beak used for picking up food. The tongue moistens the food with saliva, after which the food passes through the esophagus, or gullet, on the way to the crop, where it remains about 12 hours. The food is here softened and then passes into the stomach where it is mixed with gastric juices and passes on into the gizzard. The gizzard is the largest organ of the hen, and its office is to crush or

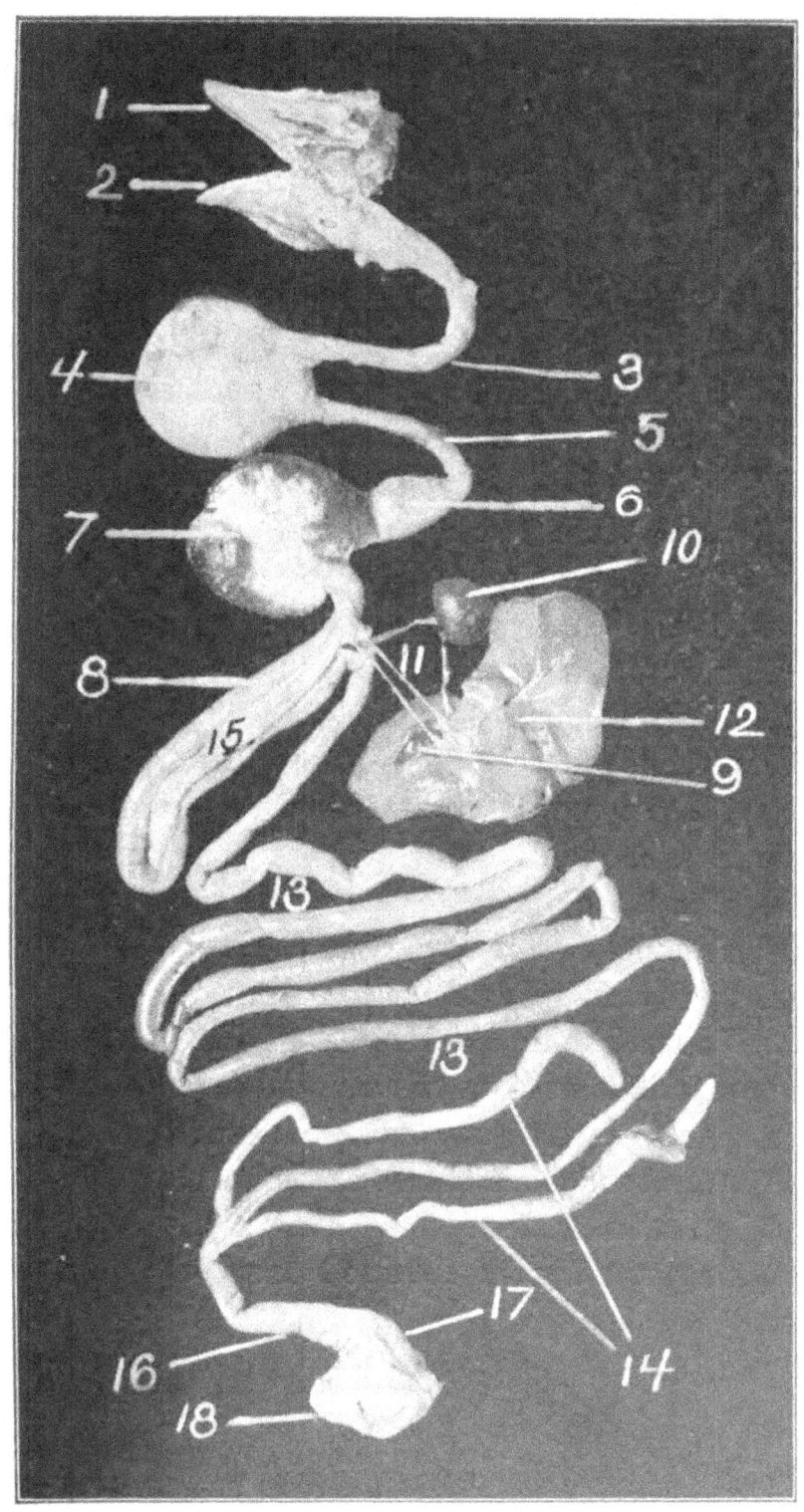

DIGESTIVE ORGANS OF THE FOWL

1, 2. Upper and lower mandibles. 3. Esophagus. 4. Crop. 5. Esophagus. 6. Stomach. 7. Gizzard. 8. Duodenum. 9. Gall bladder. 10. Spleen. 11. Bile ducts. 12. Liver. 13. Small intestine. 14. Ceca. 15. Pancreas. 16. Rectum. 17. Cloaca. 18. Anus. (Oregon Agricultural College.)

grind the food. The tough muscular walls of the gizzard aided by the grit that the hen picks up takes the place of teeth which in domestic animals grind the food. The moistened ground grain passes from the gizzard into the large intestine, or duodenum, where it is acted upon by the pancreatic juices. The bile from the liver also enters the duodenum and aids in the digestion of the fats of the food. The digestive process is here completed and the digested portions of the food are absorbed into the blood and the waste or indigestible portions forced on to the cloaca. The ceca correspond to the appendix in man, and their function is not understood. The total length of the digestive canal from beak to vent is 4 to 5 feet.

The digestive process of the fowl works with extreme rapidity. Investigations have shown that in about two days after eating, the food has entered into the making of the egg yolk. In two days after being eaten certain foods have given a color to the outer layers of egg yolk.

To keep this complex system of digestion in proper working order requires a variety of good food, abundant exercise, and fresh air in the house.

What Foods Should be Fed.—The table of composition of foods contains the names of foods that are used for poultry. This table does not, however, exhaust the list, as there are doubtless other foods that are used to a limited extent in different localities. The composition of any food not on this list may usually be obtained from the experiment stations.

CHAPTER XII

COMMON POULTRY FOODS

Among the Grain Foods wheat is more largely used for poultry than any other cereal, taking the country over. It is a safer food than most other grain foods, and there is probably no other cereal that is better relished by the fowls. It has a near competitor in corn, and whether the one or the other should be fed is largely a question of their prices. If fed exclusively on one grain, fowls would probably give better results in egg yield on wheat than on corn. Judging from the composition, wheat has a slight advantage over corn for egg production, while corn is better for fattening.

It is not a question, however, of one kind of grain; no one should expect a profit from fowls when fed one kind of food, no matter what kind of food it may be. When fed in combination with other foods it is an open question whether wheat or corn is the more economical to feed at the same price per pound for each. No serious mistake will be made by the poultryman if he makes the market price the basis for selecting wheat or corn.

Corn is an excellent poultry food. A few years ago poultry writers generally advised poultrymen not to feed it to laying hens. Chemical analysis had shown it to contain more fat-forming elements than wheat, and on this account it became very unpopular, and higher priced wheat was fed in its place. Later, however, experiment stations, in actual feeding tests, showed it to be the equal of wheat when fed in proper combinations. The Massachusetts Station secured as good, if not better, results in egg yield from corn as from wheat.

But neither wheat nor corn is a perfect ration, and other foods must be fed to "balance" it. It is a waste of food and labor to feed either wheat or corn alone.

It is an interesting fact that those states which are the largest producers of corn are the heaviest producers of poultry and eggs. This does not, however, prove the

A THRESHING SCENE
The chickens will thresh their own grain and save the threshing bill.

superiority of corn, but it disproves the old notion that corn is not a good poultry food.

Oats.—Pound for pound, oats are not worth as much for chickens as wheat or corn. Fowls do not relish oats as well as those grains. The large amount of hull on the oats is an objection. The hulls are largely indigestible. Minus the hulls, oats would be an excellent food for laying or fattening fowls. Oats are not as fattening as corn or wheat, and many poultrymen feed considerable quantities of oats to prevent the hens becoming too fat. Special care should be used in selecting oats, as they vary a good deal in quality. Only heavy, plump oats should be fed. The chief value of oats is in furnishing a necessary variety to the ration. This, of course, is true of other foods. Hulled oats, if they could be obtained at a reasonable price would be superior to corn or wheat.

Barley is not extensively fed to poultry. Chickens will not eat it if they can get wheat or corn, or, at any rate, they will eat but little of it. Where the price is not more than that of other grains, a little may be fed to give variety. Many poultry feeders use rolled or chopped barley in the mash.

Wheat Bran.—Bran is the outer covering of wheat and other grains, separated from the flour in the process of milling. Wheat bran is richer in protein than whole wheat, and has considerable ash or mineral matter other than lime. Investigations have shown bran to be low in digestibility, but nevertheless it is one of the most popular of poultry foods. There is no cereal by-product more universally used by poultry feeders than bran. Practical experience long ago demonstrated its high value for poultry, especially for egg production. For fattening it has not the same value. Its high feeding value for egg production and for growing chickens is undoubtedly due to its high mineral content, as well as protein content. It contains also more fat than either wheat or barley. These facts, added to its relative cheapness, make it an economical feed.

Middlings and Shorts.—These are other by-products of wheat that are extensively used. They have a high protein content compared with the whole wheat, and on this account and their relative cheapness make a liberal use of them in the mash desirable. Middlings and shorts are composed of the finer parts of the bran with some of the coarser parts of the flour separated in bolting.

Peas.—Where peas can be grown successfully they should be used quite extensively as a poultry food. They are richer in protein than any of our common cereals. They contain twice the quantity of protein that corn contains, and on that account are worth more pound for pound than corn or wheat.

Rye.—Rye grain is not a satisfactory poultry food. Fowls do not relish it though they eat it in small quantities. It lacks palatability. When planted in the yards in the fall it furnishes an early green food in the spring. Before the grain is fully ripe in the straw the fowls eat it more readily, and they may be allowed to thresh the grain out of the straw in the yards.

Rice.—Broken rice is used to a considerable extent in certain sections as food for small chickens. Rice polish is rich in the mineral element phosphorus.

Linseed Meal.—The meal of flaxseed from which the oil has been largely extracted in the process of manufacture of linseed oil is largely used as a poultry food. Old process meal contains more oil than the new process meal, and on that account is more valuable. Linseed meal has also a high percentage of the mineral compounds phosphorus, iron, sulphur and magnesium. It is a rich food and can only be used in limited quantities. If it can be purchased at a reasonable price, or on the basis of its protein content, it may well be used profitably as a part of a laying ration.

Buckwheat.—This is a good poultry food, but its use is limited on account of an uncertain supply and its high price in most sections.

Sunflower Seed.—The sunflower plant may be profitably used for a double purpose. It is largely used for furnishing shade. The seeds contain a high percentage of oil. They ripen about moulting time when foods of a considerable oil content are desirable. The seed may be fed in limited amount throughout the year, but during the moulting season in the growth of new feathers there is an extra demand for food of this character. It has been observed to give a glossy and attractive appearance to the plumage.

Animal Foods.—The hen is a meat eater. Animal food of some kind is necessary for fowls to maintain their health and vigor, and to make them productive either in meat or eggs. A knowledge of this fact has done more to increase the poultryman's profits than any other one thing in poultry feeding. The scarcity of eggs in winter is largely due to a lack of animal food. The fact that chickens when given the liberty of the fields in summer find animal food in the form of bugs, angleworms, grasshoppers, etc., escapes the notice of the farmer, and in winter he does not see the necessity of feeding it. In most parts of the country, during the winter, chickens are unable to obtain animal food in the fields, especially in sections where snow covers the ground. In sections with mild and open winters, they find many angleworms, especially during the rainy season. But in most sections, if not in all, fowls must be liberally fed with some kind of animal food to obtain best results.

There are a number of forms in which animal food may be fed. Fresh, lean meat is undoubtedly the best kind of animal food. It is the lean meat that furnishes the protein, but there is no objection to having the lean mixed with a little fat; this may be an advantage at times. *Fresh meat scraps* or *cut bone* from the butchers' stalls are an excellent egg-maker. Some butchers keep a bone cutter and sell the meat and bones all ready ground or cut up. When one has a sufficient number of hens, say 25 or more, it will pay to buy a good bone cutter and cut the bones. The scraps contain a large proportion of bone, and the fowls eat these very greedily, as well as the meat. They furnish the mineral matter necessary for bone making and for eggshell making.

Skim milk will take the place of animal food if fed

liberally enough. The trouble with skim milk is that it is not concentrated enough; that is, it is largely water, 90 pounds in a hundred being water. In other words, in 100 pounds skim milk there are only 10 pounds food. Even with milk kept before them all the time to drink, laying hens will not get enough of it to supply the demand for animal food. If wet mashes are fed, by using skim milk to mix the mash they will get more of it in this way. By feeding it clabbered the fowls will get more food out of it. Probably the best way to feed milk is to make "cottage cheese" out of it. This is a splendid food when properly made. In that form fowls will consume enough to supply the demand for animal food.

It is made in this way: Set a can of skim milk in a place having a temperature of 75 to 80 degrees. In 18 to 24 hours the milk will coagulate (thicken). Then break up into pieces the size of large peas or smaller; set can in a pail of hot water, stirring the curd until a temperature of 90 to 95 degrees is reached; hold at this temperature for 15 or 20 minutes, without stirring. Then pour the contents of the can into a cotton sack and hang up where the whey can drain off. The milk should not be boiled. Salt it a little. It will keep a day or two.

Buttermilk is largely used in fattening poultry, the large fattening establishments using it generally for mixing the ground grain. In the feeding of small chicks it has special value as a preventive of white diarrhœa. It is also profitably used in the laying ration. The mash may be mixed with it and the fowls also given all they will drink of it. At the Ontario Agricultural College, Professor Graham has used it successfully as a substitute for other forms of animal food, and also as substitute for water. Sour milk has much the same value as buttermilk. By withholding water more buttermilk is taken by the fowls.

Unless they can be made to use large quantities of it, enough of the animal nutrients will not be secured to supply the need for animal food. In 100 pounds of skim milk or buttermilk there are only about 10 pounds of solids or food, and this should be considered in arriving at an estimate of its value.

Whey also may be used as a source of animal food, but as may be seen in the table of composition of foods, it has a lower value than skim milk and buttermilk.

Milk Albumin.—This is a by-product of the manufacture of milk sugar. It contains little moisture and a high percentage of protein, but it is low in other nutrients. All forms of milk foods lack in mineral matter, also in fat. Where milk is used bones should be fed either dry or green to furnish the required mineral matter.

Beef Scrap is the most convenient form in which to feed animal food. This is a by-product of the large packing houses, and contains meat and bones in varying proportions which have gone through a boiling and drying process. It contains, therefore, little moisture compared with fresh meat scraps. It varies considerably in composition, but should contain from 50% to 60% protein. Beef scrap varies also in quality. It should be light colored with a meaty flavor and somewhat oily to the touch. When boiling water is poured over it, it should have a fresh, meaty flavor. If it gives off a putrid odor, do not feed it.

Fish Scrap is coming into use as a substitute for beef scrap. Its practical value, however, compared with beef scrap has not been experimentally determined, but the practice of feeding it is growing, especially on the Pacific Coast. The oil being largely removed in its manufacture, there is no fishy taste transmitted to the eggs and chickens by its use. If fresh fish, however, is liberally eaten there will be a distinct flavor given to the egg.

Green Foods.—Green food of some kind is an essential part of the ration or diet. The health of the fowls and the demands of egg production require it. The lack of a sufficient supply of green food is one cause of the scarcity of eggs in winter. During the summer the farmers' flocks, which furnish the markets with the large proportion of eggs and poultry, usually find all the green food necessary, but in winter, since the farmer does not realize the importance of providing green food, the chickens do without it and we do without the eggs. Spring is the natural laying season; but by seeing to it that the fowls get the same kind of food in winter that they do in spring or summer, it is possible to overcome largely the egg famine in winter. Fowls should have all the green food they will eat at all times. Green food is cheap, or should be grown cheaply with good management.

Green food may be fed in different forms. *Clover* or *alfalfa* or grass in the fields; clover leaves or alfalfa leaves in the haymow or in the haystack, make excellent green food; vetch, peavine, rape, rye, kale, mangels, sugar beets, cabbages, lettuce or turnips will fill the bill. It will be noticed that these green foods have a larger percentage of mineral matter or ash, and of protein, than the grain foods. Alfalfa and kale are especially rich in protein and ash. Clover, alfalfa, grass, rape, kale and vetch, will give good color to the yolk of the egg; beets will not. Alfalfa and clover will give eggs of good quality and flavor. Kale, cabbages and rape will give a slightly undesirable flavor to the eggs if eaten heavily, but not enough to injure their selling value materially, if at all. If fed regularly, however, so the fowls may eat it at will, there is no evidence that an undesirable flavor will be imparted to the egg.

In western Oregon and the Pacific Coast generally *thousand-headed kale* is probably the most profitable crop to grow for winter forage. Here it grows to perfection, and an acre may be made to produce 40 tons of green forage. For winter green food, kale is transplanted in July from seed sown in May or June. For summer forage it is planted early in the season. It is possible in western Oregon to have green kale the year round. For a flock of one hundred hens, about two hundred plants will furnish green food enough for a year where the soil has plenty of fertility and moisture. The plants should average 20 pounds each. The chickens will eat about half the weight of the plant, the balance being stalk which they do not use. Cattle will eat most of the stalk. Planted in July, the kale may be fed from October to April. Planted early in the spring from seed sown in the fall, it will be ready for use in the summer. In the early part of the season the lower leaves may be stripped off and the rest of the plant will continue to grow.

The plants are set about 3 feet apart each way. A very small piece of ground, therefore, will grow enough kale for one hundred hens. A strip of good land 16 feet wide and 100 feet long should furnish enough green feed in the form of kale for one hundred hens. At that rate, an acre of kale will furnish green food for 2,000 hens throughout the year. Kale may also be utilized for shade for fowls. Where fowls are yarded, by having double yards, it is possible where kale grows the year around to make it furnish the green food and shade all the year. Kale will keep the yards in sanitary condition, turning the manure and filth into a revenue.

Vetch and Oats.—This makes a good combination for early spring green feed in sections where vetch grows well. Vetch is a leguminous crop, like clover and alfalfa.

Beets.—Sugar beets and mangel-wurzels are used by many poultrymen for green food. The tops may be fed green and the beets stored for winter use. One peculiarity of beets is that they do not furnish the coloring matter for the egg yolk, as do clover, alfalfa, kale, and other greens. In case the yolk is too highly colored, beets may be substituted for part of the other green feed that is responsible for the color.

Beet Pulp.—Dried beet pulp is now used to a considerable extent in stock feeding. It may be used as green food for poultry. In addition to its value as a succulent food, it contains a fairly high percentage of mineral matter. This makes it of more value than some other green foods. There is little authoritative data on the subject of beet pulp as a poultry feed, and at the present time it should be used experimentally.

Sprouted Oats.—Sprouted oats may be resorted to where other forms of green feed are not available. This green food is very greatly relished by the fowls.

Oats and Peas.—"Oats and peas sown together very thinly, with a liberal seeding of red clover and a very little rape, make a good combination. The oats and peas furnish a rapid growth of green food, a good deal of which will get tramped down and some will go to seed, but it will serve to protect the clover and rape, which will make good food for the late summer and fall pasturage. Three pecks of oats, two pecks of peas, one pint of rape seed and five quarts of red clover seed will be a good proportion for seeding. The oats and peas should first be harrowed in deeply, then the clover and rape seed should be mixed and sown, then lightly scratched in with a weeder."—PROF. JAMES E. RICE.

Potatoes may sometimes be fed for variety, if boiled and mixed with mash, but they are not a good egg food; they are better fitted for fattening.

Cabbages are very much relished. Apples of sour varieties should be sparingly fed to poultry. On the whole, clover and alfalfa are probably the most satisfactory green food we have. In coast regions, where it grows throughout the year, the thousand-headed kale by reason of its heavy yielding quality is probably the most profitable green food to grow. But it may be supplemented by other green food such as clover, alfalfa or lawn clippings.

Grit.—"The hen coins silver out of sand." The chickens need grit as well as the poultryman, but of a different kind. There are two views about chicken grit, and I do not pretend to reconcile them. One view is that the chief function of grit is to grind the food; the other is that grit itself is food. Whatever the function, we know that grit is a necessary part of the diet, and the health and productiveness of the fowls require a liberal consumption of grit. On most farms, where the fowls have the liberty of the fields, they will pick up all the grit necessary, but on soils having little or no sand or gravel, and where the fowls are confined in yards, it is absolutely necessary to furnish grit just as regularly as food. With a gravel bed located near the poultry yards, the grit question is easily and cheaply solved. Give them plenty of sharp gravel and sand to work over. Where this is not available, grit may be cheaply purchased at the poultry supply houses. Keep it where the hens can get it at any time.

Egg-shell Material.—Ordinary grit probably furnishes material for egg shells, but in addition it will be found advisable to feed special shell material. The grains do not contain lime enough to furnish sufficient shell material for heavy laying hens. Ordinary sea shells and especially oyster shells are largely used for this purpose. They are very readily dissolved in the gizzard. The egg-eating habit among hens is sometimes acquired because of a scarcity of lime or shell material in the ration.

Charcoal is a bowel regulator, and most of the successful poultrymen feed it regularly. It may be kept in a box or hopper where the fowls can eat it at will. *Salt* is an aid to digestion. It may be fed at the rate of about an ounce or two ounces per day to one hundred hens.

Pepper is stimulating and should not be fed except in very small amounts. Hens in good health do not need it. It is sometimes useful in case of sickness in the flock. If the flock should be afflicted with colds a little red pepper may be mixed in the soft feed.

CHAPTER XIII

METHODS OF FEEDING

While a knowledge of the composition of foods should be possessed in order to feed successfully, it is equally important that there should be a knowledge of how to feed. It is not sufficient that the poultryman should have all the best available poultry foods. He may have all the necessary foods, and fail in the purpose for which he feeds. The laying hens may have all the best available foods and yet refuse to lay eggs unless the food comes to them in a certain way. Success in feeding for egg production will be measured largely by the methods followed in feeding.

Exercise and Activity.—The secret, if there be any secret, in how to feed to get eggs is to feed in such a way that the natural activity of the hen may be maintained. In the production of flesh or meat in domestic animals as well as in poultry, activity or exercise counts for little, nor is exercise so important for the cow that is producing milk, but activity is the life of the hen. She is given toe-nails to scratch with, legs to walk with, wings to fly with. If there is any one characteristic more than another that indicates the good layer, it is the active use of those organs in her every-day life.

The vigor of the hen comes largely from her activity, and it is the vigorous hen that lays. The reason hens on free range often do better than others confined in yards, is largely because of the active life they live. Under the free-range system the poultryman need concern himself little on this point, but when fowls are confined in yards, which is an artificial condition, great care must be taken to furnish the exercise or the incentive to exercise. A hen that "stands around" all day, only exerting herself enough to eat out of a hopper, is an unproductive hen.

The exercise is best furnished by providing a roomy scratching floor or shed covered with a deep litter of straw. This may be from 8 to 12 inches deep, and should be kept reasonably dry. The whole grain food should be scattered in this straw. There will be no waste in this, as the fowls will find about every kernel. The skill of the poultryman comes in feeding enough at a time, without having to feed too often, to keep the hens busy at work a large portion of the day. If too much is given at a feed the fowls will soon satisfy their appetites, while if too little is given they will soon clean it up and there will be nothing to scratch for. It is not necessary to keep them scratching all day. Leghorns, for instance, will do nearly as well when fed in a hopper or box. If they have a yard and a floor they will exercise themselves whether compelled to dig for their food or not. Forced exercise, however, is necessary for the larger or less active breeds.

In an experiment three pullets kept in a small pen on a board floor without any litter, laid 116 eggs in a year, an average of 38 2-3 eggs each. One of these was a Leghorn pullet which laid 52 eggs. Leghorns fed in straw averaged 169 per fowl, and others fed in boxes or hoppers averaged 161. Both were kept in pens without floors and had access to an outside yard. They exercised a good deal by scratching in the earth. Two pens of Plymouth Rocks averaged 141 fed in straw, and two fed in boxes averaged 118 eggs each. In each case the ration was the same. It is seen that the method of feeding was responsible for a variation in yield of from 38 eggs per fowl to 169. The experiment showed that no exercise, or forced idleness, was ruinous both to production and to health of fowls. Second, it showed that Leghorns, or the active breeds, will do well even though they are not forced to scratch; but that the heavier breeds need some "forced" exercise.

Feeding yarded fowls in the litter, therefore, is a decided advantage with some breeds, and it is an advantage with any breed. A Leghorn given the liberty of a yard and a floor to scratch on, even though all grain be fed in a hopper or box, will take exercise enough to produce fairly well. The chief disadvantage of feeding in the litter is that the grain is liable to become contaminated with the droppings of the fowls, which is a fruitful method of carrying disease from one fowl to another. This method, however, is usually necessary with most fowls, and with care in renewing the straw often enough, little danger need be feared from this source. The droppings from the fowls at night should not be permitted to mingle with the litter.

Ground or Unground Grain.—It pays to feed part of the grain ground. It is a saving of energy, and energy is furnished by the food; therefore, it will save food to grind some of the grain for the fowls. Ground food is more quickly digested and assimilated than whole. The hen can manufacture the eggs faster with ground food than with whole grain. Experiments by Wheeler showed that fowls having half their grain ground and moistened required 20% less food to produce a dozen eggs than fowls having all whole grain. Fowls, however, relish the whole grain, or a large percentage of it whole. Probably one-third of the grain ground would be a safe limit to feed. The danger in feeding one-half or more of it ground would be that the fowls would be liable to lose appetite and not eat enough to fill the demand for heavy egg yield.

Best Time to Feed Wet Mash.—If fed heavily on wet mash in the morning, the fowls would gorge themselves and would not be as active the rest of the day as if fed a light feed of grain in the litter in the morning. A good feed of mash about an hour before going to roost, followed by a feed of whole grain, will give satisfactory results. In cold weather especially the practice of feeding whole grain liberally the last feed of the day is a good one. Whole grain will "stay with them" better throughout the long, cold night than mash, and keep up the heat of the body better. It will save feeding in the morning if at the last feed at night enough grain is thrown on the litter to more than satisfy the fowls, and leave some for them to begin scratching for in the morning. Where wet mash is fed the first thing in the morning, this should not be done. The writer prefers to feed the mash in the morning, just as soon as the fowls come from the roost, but to feed only as much as they will eat up readily so they will go to work scratching in the straw for the whole grain. It is not so material at what time of the day the soft food is fed, as it is that the fowls be kept active and retain their appetites.

Length of Day and Egg Yield.—There is no doubt some connection between the lower egg yield in winter and the shorter days. When the spring comes and the feeding day lengthens there is an increase in production. Some of this increase is probably due to the longer period of activity and the necessarily greater consumption of food. Some support is given to this theory by recent private experiments in the use of electric light in the poultry house. It is a point worthy of further investigation.

Wet versus Dry Mash.—Dry feeding saves labor. Fowls relish the wet mash better. Wet mash economizes in the ration. By feeding the mash dry, it may be fed once a week in hoppers. When fed moist it must be fed once a day. Fowls will eat wet mash more greedily than dry, and for that reason more care is required in feeding it. If given too much, they will gorge themselves and stand around lazily most of the day; this should be guarded against. Where skim milk is available it is possible to cheapen the ration by feeding wet mash. Cheap by-products, such as bran and middlings, may be made to make up a large proportion of the ration by mixing them with milk. By making a mash with milk, more milk may be fed to the fowls. It will also cheapen the ration where skim milk is cheap by saving on higher-priced animal foods. Where heavy feeding of ground grain is desired, it should be fed wet. On the majority of the large poultry ranches of the Petaluma, Cal., and of the Little Compton, R. I., districts the wet mash method is used.

When skillfully fed, the wet mash will give better results in egg yield than dry. The high egg records of the Oregon Station were secured by wet mash feeding. Results of experiments by Rice are slightly in favor of dry mash. Gowell also secured results favorable to dry feeding. In mixing wet mash, enough water or milk should be used to make the mash crumbly. It should not be sloppy. Usually about as much ground grain, by weight, as milk or water will be about right.

The results in feeding mash do not depend upon the moisture or lack of moisture in it, but upon the amount of ground grain consumed. It matters little whether the water is put into it by the feeder, or whether the hen herself drinks the water from the creek or the water fountain.

Feeding Dry Mash.—The dry mash is fed in hoppers large enough for a week's supply or more, and the fowls allowed to eat it at will. The dry mash may have the same composition as the dry material in the wet, but about 10% of its weight should be beef scrap. The fowls will eat it more readily then. Without the beef scrap they will not eat enough of the ground grain. In addition a hopper of beef scrap may be kept before them all the time. This will insure that they get enough of the animal food.

Cut Bones may be fed every day, or three times a week, as much as the fowls will clean up in 15 minutes. Three to four ounces per hen per week is about right. More will be consumed during heavy laying than at other times.

Cooking Food.—It does not pay usually to cook feeds. Most feeds give better results when fed raw. Starchy feeds, such as potatoes, are improved by cooking, but usually it is better not to cook feeds. In feeding raw meat foods, there is some danger of the fowls contracting disease. If liver or lights are fed, they should be boiled to kill any disease germs there may be in them. Digestion experiments at Geneva (New York Report, 1885), show that the digestibility of the protein in several of the common stock feeds was injured by cooking.

Hopper Feeding.—There are two fundamental considerations in methods of feeding. The first is the method of weighing out at each feeding a certain definite amount of feed. The second allows the hen herself to make good from the hopper any lack of nutrients of any particular kind. The writer believes it imperative that the hen be allowed considerable latitude in satisfying her wants and in making good any shortage of at least the mineral and animal feed in the ration. It is not conceivable that in a flock of one hundred hens where the individual egg production varies, as we know it does, the same amount and kind of feed will satisfy all of them. The heavy producer requires more of the animal protein foods and more of the mineral, and the only practicable method is to furnish those nutrients *ad libitum* to the flock.

No Hard and Fast Rules.—In what has gone before the attempt has been made to give to the reader in concise form information in regard to the general principles of feeding,

METHODS OF FEEDING

OREGON STATION OUTDOOR DRY FOOD HOPPER

This hopper has four divisions for different foods. The fowls do not pull the feed out of it onto the ground. (Designed by C. C. Lamb.)

and the composition and values of various foods. It is not presumed to lay down any hard and fast rules which must be followed by the poultry feeder.

The Food Requirements vary and methods of feeding vary in different sections of the country and even on different farms in the same section. A large latitude must be allowed the individual farmer or poultry-keeper. The highest success will not be attained where the poultryman is content to follow set rules and blindly attempt to make his conditions and environment conform to the feed rations rather than make the rations conform to his special conditions. Having a knowledge of foods and principles of feeding, and the

OREGON STATIONARY OUTDOOR DRY FOOD HOPPER

Showing inside construction.

food requirements of the fowl, he is master of the situation and will be able to formulate rations that will give him the most profitable returns.

The Price of Foods will largely govern choice of a ration. Profitable poultry production is not a question of the best foods any more than it is a question of the cheapest foods. That is to say, the best foods from the standpoint of composition and palatability may produce more eggs or more meat but may produce less profit than other foods that are not so valuable, pound for pound, on account of their lower cost. The feed bill may be so high that the poultryman is robbed of his profits. No one kind of food is so essential that the poultryman must feed it no matter what its price may be. If this one fact were thoroughly understood and acted upon it would save probably millions of dollars to the poultry-keepers of the country.

Rations.—With this understanding a few sample rations for egg production are here given. The weights of feed are in pounds, and are figured on the basis of one average hen for one year. It will be understood that these amounts will vary, first, as the size of hen varies; second, as production varies, and third, as the climate or temperature varies. The amounts given approximate closely the amounts required in egg production. The safe rule to follow is to increase or decrease these amounts daily as demanded by the fowls. There must be no stinting of food if a steady production of eggs is to be maintained.

The choice of animal food is left to the feeder, 50 pounds skim milk or buttermilk, 10 pounds cut bones, and 5 pounds beef scrap being estimated as of about equal value. The same is true of green food, 15 pounds of green alfalfa or clover being equal in value to 20 pounds kale. This does not exhaust the list of animal food nor of green food. It may be, for example, that fish scrap is more available in some sections, and various kinds of green feed may be fed with satisfactory results. Such foods are discussed in another place.

SAMPLE RATIONS PER HEN PER YEAR (IN POUNDS)

Number of Ration

	One	Two	Three	Four	Five
Wheat	..	60	40	30	20
Corn	60	..	..	10	20
Oats	..	..	10	10	10
Bran	..	..	10	10	10
Middlings	..	..	5	5	5
Linseed Meal	..	..	..	5	5
Skim Milk, Buttermilk (with dry bone)	50	50	50	50	50
Cut Bone	10	10	10	10	10
Beef Scrap	5	5	5	5	5
Vetch, Alfalfa, Clover	15	15	15	15	15
Kale	20	20	20	20	20
Oyster Shell	3	3	3	3	3
Salt	¼	¼	¼	¼	¼

COMPOSITION OF RATION 4 (NOT INCLUDING GREEN FOOD AND SHELL)

Water	8.12	pounds
Ash	2.14	"
Protein	13.32	"
Carbohydrates	48.33	"
Fat	3.09	"
Total	75.00	"

Ratio of protein to carbohydrates and fat 1: 4.14

It is estimated that 30 pounds milk is about all that a hen will ordinarily consume in a year. If no water is given, the fowls will use a great deal more milk or buttermilk, probably enough to supply the full demand for animal food.

Under farm conditions, however, where fowls have free range and find a good deal of animal food in the fields, 30 pounds should be sufficient. They should have access to it at all times. The amount they will consume will be governed in part by the amount of insects found in the fields. By making the milk into cottage cheese and feeding the fowls all they will eat of it, they will get all the animal food required. Milk when closely skimmed has very little fat, while bones and beef scrap have a large amount of fat; it can, therefore, be fed to advantage in rations that in other respects are richer in fat than would be necessary or advisable where cut bones are fed. Good fresh cut bones fed regularly will give better results than either milk or beef scrap, but the cost is sometimes prohibitory, and there is danger of the meat not being fresh. Where milk is used as animal food it should be supplemented with dry or ground bone that will furnish the necessary mineral matter that is lacking in the milk.

It is not very material what kind of green food is fed. The important thing is to give the fowls all they will eat. Alfalfa and clover have about equal feeding value. In winter, alfalfa and clover leaves make good green food. Kale has a higher percentage of water than green alfalfa or clover or vetch.

The table gives five rations, numbered from 1 to 5. No. 1 is rated as the poorest and No. 5 the best. Corn is the only grain in ration 1. In No. 2 wheat is fed in place of corn, but in other respects they are the same. Number 2 is placed ahead of No. 1 because it has slightly more protein. Both of them are deficient in the egg-making material, protein. Though not an ideal ration by any means, either 1 or 2 would be an improvement on many rations fed on the farms, but for heavy egg production neither has enough protein. No. 3 is better than Nos. 1 or 2 because it has a variety of grains and a little more protein. Nos. 4 and 5

METHODS OF FEEDING

should give a heavy egg yield if properly fed. They are equal in protein, but No. 5 has more fat than 4. Corn, which has more fat than wheat, should be fed more liberally during the cold weather than during the summer. Ration 5 therefore should be a better winter ration than 4.

How to Feed the Rations.—To get the best results from rations 1 and 2 the fowls should have free range on the farm. These rations would be altogether impracticable for yarded fowls. A light feed of corn or wheat should be given in the morning, and all they will eat up at night. If the fields contain bugs and worms and other animal food they will get exercise hunting and scratching. There will be weed seeds and waste grains of different kinds at different seasons and these will give them incentive to exercise, and at the same time help to balance the ration. Under such conditions it would be possible to secure a fairly good egg yield from rations 1 and 2.

But where other grains may be secured it would be a serious mistake to confine the feeding to such rations. These two rations may be very much improved by the simple method of keeping a hopper of dry bran accessible to the fowls at all times. They would be further improved by adding a little middlings or shorts and a small amount of linseed meal to the bran in the hopper. This would give us ration No. 3. Adding the bran, middlings and linseed would cut down the amount of wheat necessary. This makes a very good ration for the general farm. It is practically a balanced ration, at any rate it gives the hen the opportunity to balance her ration; besides it requires very little labor in the feeding. If cheaper than wheat, corn may be substituted for wheat.

Ration No. 4 is an improvement on ration 3. Ten pounds of corn is substituted for 10 pounds of wheat. Even if corn costs a few cents more per bushel than wheat, it will pay to feed this quantity. Ration 5 is an improvement on No. 4. If corn is as cheap or cheaper than wheat this ration should be fed.

The Mash Feeding.—If it is desired to feed a dry mash, the bran, middlings and linseed should be put together in a hopper where the fowls can help themselves at will. The hopper should never be empty. It will improve the dry mash still further if beef scrap be added, using from 10 to 15 pounds in 100 pounds mash. This will induce the fowls to eat more of the dry mixture. Where milk constitutes the animal food, it will be better to use a soft mash, mixing it with milk; also keeping milk where the fowls can drink it whenever they want it. If no milk is available and the mash is moistened with water, a hopper of beef scrap should be supplied. In place of beef scrap, cut bones may be fed. There is no danger in the fowls eating too much beef scrap, assuming of course that its quality is good. The only danger is in permitting the hopper to get empty, for, after being without animal feed for a few days, they will eat too much of it when it is given to them again.

Five pounds beef scrap is given as the amount necessary for an average laying hen for a year. This amount will vary with different hens. The fowls may not eat 2 or 3 pounds, or they may eat 6 or 7, but it is safe to permit each hen to eat just what she requires.

Oregon Station Method.—Ration No. 4 is practically the one used at the Oregon Station. The mash is fed moist. Sour milk or buttermilk is used in mixing it, a little more milk than ground grain being used.

On account of the high prices of corn some years, less of it has been used than is shown in table. Unless corn gets down to about the price of wheat, the corn that is fed is ground and put in the mash. The whole grains are then wheat and oats.

METHODS OF FEEDING

The mash used during the year 1912-13 was as follows by weight:

Bran	4 parts
Middlings	1 part
Ground Barley	1 part
Ground Corn	1 part
Linseed Meal	½ part
Milk	8 parts

Salt is added at the rate of about 4 ounces per hen per year. The proportion of bran is reduced in case the droppings show a watery condition.

The mash is thoroughly mixed and fed as soon in the morning as the fowls are ready to eat and before they have had anything else. The amount of mash, dry material, averages about one ounce per hen per day. The amount fed does not vary very much from morning to morning, but if there should be any left in the trough for more than an hour after feeding, it is taken away and next day less is fed. Then the amount is increased as their appetite for mash increases, until they are getting the normal amount. The idea is to get them to eat as much as possible in about an hour.

The Skill of the Feeder comes in largely in so feeding that the fowls will eat the required amount of the mash of ground grains. A heavy laying hen requires a full crop of grain at night, but there should not be any whole grain left over night for them to eat in the morning, otherwise enough mash will not be eaten.

Feeding the Oats.—About ten to eleven o'clock a feed of oats is given. This is thrown in the litter, just enough to keep the fowls busy scratching for an hour or two.

Feeding the Wheat.—In the afternoon or evening, or between two and three o'clock during the short days, between three and four in the long days, whole wheat is fed in the litter, as much as the fowls will clean up before going to roost, and they must have as much as they will eat, and no more.

Beef Scrap is kept in a hopper all the time, care being taken that the hopper never gets empty before being refilled. In addition, fresh cut bone is fed three times a week, about an ounce per week per fowl.

Oyster Shell, Charcoal and Grit are also kept in separate hoppers.

Green Food is before the fowls all the time. It is usually kale. They help themselves at will. A head of kale is hung up fresh in the morning, and they pick at it whenever they want it. This is supplemented at different seasons by vetch, clover, and other green stuffs that grow in the yards. The fowls are changed twice a year to clean ground and green stuff is growing in the yard when the fowls are put into it.

Cleanliness.—It is important that feeding troughs and drinking vessels be kept clean. They should be scalded frequently with boiling water. Do not throw feed on dirty, filthy ground.

Changing the Ration.—Radical changes in the ration should be avoided. The feeder should first map out his system of feeding and stay by it. Remember that the food is not everything, and when the fowls are not laying do not conclude that it is the fault of the ration unless you have definite knowledge that it is. A sudden change to new food, even though the new food may be better than the old, will check egg production for a considerable time. If changes are to be made, it is better to make them gradually.

Regularity.—Stated times should be given to the feeding. A "feast and a starve" will not satisfy the laying hen. During the winter the hen should go to roost with a full supper to sustain her through the long night, and just as early as she can see to eat in the morning her breakfast should be ready.

Summing Up.—Feed wholesome food; feed liberally; feed regularly; feed a variety. After that, the only secret in feeding is to feed *activity* into the hen.

Cornell Rations for Laying Hens.—The following whole grain mixture is fed morning and afternoon in a straw litter:

By weight Winter	By weight Summer
60 lbs. wheat	60 lbs. wheat
60 lbs. corn	60 lbs. corn
30 lbs. oats	30 lbs. oats
30 lbs. buckwheat	

The following mash is fed dry in a hopper kept open during the afternoon only:

By Weight Winter and Summer	By Measure Winter and Summer
60 lbs. corn meal	57 qts. corn meal
60 lbs. wheat middlings	71 qts. wheat middlings
30 lbs. wheat bran	57 qts. wheat bran
10 lbs. alfalfa meal	20 qts. alfalfa meal
10 lbs. oil meal	8 qts. oil meal
50 lbs. beef scrap	43 qts. beef scrap
1 lb. salt	½ qt. salt

The fowls should eat about one-half as much mash by weight as whole grain. Regulate the proportion of grain and ground feed by giving a light feeding of grain in the morning and about all they will consume at the afternoon feeding (in time to find grain before dark). In the case of pullets or fowls in heavy laying, restrict both night and morning feeding to induce heavy eating of dry mash, especially in the case of hens. This ration should be supplemented with beets, cabbage, sprouted oats, green clover or other succulent food, unless running on grass-covered range. Grit, cracked oyster shell and charcoal should be accessible at all times. Green food should not be fed in a frozen condition. All feed and litter used should be strictly sweet, clean and free from mustiness, mould or decay.

AVERAGE COMPOSITION OF EGGS, EGG PRODUCTS AND CERTAIN OTHER FOODS

	Refuse, per cent.	Water, per cent.	Protein, per cent.	Fat, per cent.	Carbohydrates, per cent.	Ash, per cent.	Fuel value per pound calories
Hen:							
Whole egg as purchased	11.2	65.5	11.9	9.3	....	0.9	635
Whole egg, edible portion	....	73.7	13.4	10.5	....	1	720
White	....	86.2	12.3	0.2	....	0.6	250
Yolk	....	49.5	15.7	33.3	....	1.1	1,705
Whole egg boiled, edible portion	....	73.3	13.2	12	....	0.8	765
Evaporated hen's egg	....	6.4	46.9	36	7.1	3.6	2,525
Cheese as purchased	....	34.2	25.9	33.7	2.4	3.8	1,950
Sirloin steak as purchased	12.8	54	16.5	16.1	....	0.9	985
Sirloin steak, edible portion	....	61.9	18.9	18.5	....	1	1,130
Milk	....	87	3.3	4	5	0.7	325
Oysters in shell as purchased	81.4	16.1	1.2	0.2	0.7	0.4	45
Oysters, edible portion	....	86.9	6.2	1.2	3.7	2	235
Wheat flour	....	12	11.4	1	75.1	0.5	1,650
Potatoes, as purchased	20	62.6	1.8	0.1	14.7	0.8	310
Potatoes, edible portion	....	78.3	2.2	0.1	18.4	1	385

FEEDING SMALL CHICKENS

Different rations may be successfully fed to chicks. The following have been tried and are recommended by the respective stations:

Oregon Station Ration

Starting food	Grain mixture	Mash mixture
Bran mixed crumbly with soft-boiled egg; or stale bread squeezed dry out of milk.	1 lb. cracked wheat 1 lb. cracked corn	3 lbs. wheat bran 1 lb. wheat middlings or shorts 1 lb. corn meal Pinch of salt added when mixing

FIRST FEEDING TIME (24 TO 36 HOURS OF AGE)

First Week.—Starting food twice a day; grain mixture three times a day on clean sand; after two or three days, grain in litter; clean water; grit, charcoal, cracked bone, in separate dishes; green food.

One to Three Weeks.—One feed a day of moist mash, what they will clean up in an hour; grain mixture in litter two or three times a day; grit, charcoal, cracked bone, and beef scrap in hoppers; water; green food.

Three to Six Weeks.—Morning feed of moist mash; two feeds of grain mixture; dry middlings in a hopper, if signs of diarrhœa appear; hopper-fed beef scrap; water, grit, charcoal, cracked bone, always available; milk to drink; green food.

After Six Weeks or On Range.—Morning meal of moist mash; two feeds of grain mixture; milk (or beef scrap), charcoal, grit, bone, water. Oats may be added to the grain mixture, if desired; the proportion of wheat may be increased or decreased as it becomes lower or higher in price than corn.

Cornell Ration

Starting food	*Grain mixture*	*Mash mixture*
8 lbs. rolled oats	3 lbs. wheat	3 lbs. wheat bran
8 lbs. bread crumbs	2 lbs. corn	3 lbs. wheat middlings
2 lbs. sifted beef scrap	1 lb. hulled oats	3 lbs. corn meal
1 lb. bone meal	Fine cracked for the youngest chicks; whole wheat and hulled oats and larger cracked corn for older chicks; oats omitted for range chicks.	3 lbs. beef scrap
Moistened with skim milk.		1 lb. bone meal
		Fed dry from first meal; moist, and dry after five days.

FIRST FEEDING TIME (36 TO 48 HOURS)

First Five Days.—Starting food five times a day, what they will eat in 15 minutes; grain mixture in tray of dry mash always available; fine grit, charcoal, bone, and green food scattered over other food; water.

After Five Days.—Grain twice a day in litter; scanty feed of moist mash three times a day; as chicks grow older, two feeds of moist mash, then only one—at noon; water, grit, charcoal, cracked bone, always at hand, and hopper-fed beef scrap if desired; milk to drink. Chicks should be hungry once a day, preferably in the morning.

On Range.—Grain, dry mash, beef scrap, grit, shell, bone, water, always at hand. One meal of moist mash if desired.

Maine Station Method

Starting Food	*Grain Mixture*
4 lbs. wheat bran	15 lbs. cracked wheat
3½ lbs. corn meal	10 lbs. pinhead oatmeal
2 lbs. screened beef scrap	15 lbs. fine cracked corn
1 lb. alfalfa meal	3 lbs. fine cracked peas
½ lb. linseed meal	2 lbs. broken rice
	5 lbs. chick grit
	2 lbs. charcoal

METHODS OF FEEDING

Mash Mixture No. 1
- 2 lbs. wheat bran
- 3 lbs. corn meal
- 1 lb. Daisy flour (or other low-grade flour)
- 1 lb. screened beef scrap
- ½ lb. linseed meal

Mash Mixture No. 2
- 1 lb. wheat bran
- 2 lbs. corn meal
- 1 lb. wheat middlings
- 1 lb. beef scrap

FIRST FEEDING TIME (36 TO 48 HOURS)

To Three Weeks.—Two feeds of starting food, scalded and mixed with rolled oats, two parts of oats to six of mixture; two feeds of grain mixture in light litter; green food; fine grit, charcoal, cracked bone, and clean water always before the chicks.

Three to Six Weeks.—Substitute mash mixture No. 1 (moist) for the starting food; otherwise as above.

On Range.—(After six or eight weeks.) Constant supply of wheat, cracked corn, beef scrap, cracked bone, oyster shell, and grit in separate troughs or hoppers; hopper-fed mash mixture No. 2.

Ontario Agricultural College Ration

Starting Food
- 4 lbs. bread crumbs
- 1 lb. hard boiled egg
- Fed dry

Grain Mixture
- 30 lbs. cracked wheat
- 30 lbs. granulated oatmeal
- 30 lbs. fine cracked corn.
- 10 lbs. small grit

Mash Mixture
- 10 lbs. wheat bran
- 10 lbs. shorts
- 10 lbs. corn meal
- 3 lbs. animal meal

FIRST FEEDING TIME (24 TO 48 HOURS)

First Two Days.—Starting food, fed five times a day; lukewarm water to drink.

After Two Days.—Three feeds of grain mixture, with one of bread and milk, and one of whole wheat; or with two feeds of moist mash; fresh boiled liver twice a week, if obtainable—in that case, animal meal omitted from the mash; for chicks on range with the hens, the grain mixture may be hopper-fed.

After Eight Weeks.—Moist mash in the morning; grain noon and night. An increase in the proportion of animal food will hasten the development of the chicks.

FEEDING AND MANAGING THE GROWING STOCK

Food requirements vary according to stage of maturity. The fact that most fowls have more or less free range and are able to find much natural food, which helps to supply any lack of nutrients in the ration fed them, in other words enables them to balance their ration, lessens the importance of varying the feeding according to special needs of production. It is true, nevertheless, that the food requirements are very different for the chick from the shell to the end of the brooding period, and from the end of the brooding period to maturity; also for the laying hen and the developing pullet and for the pullet and old fowls. The small growing chick must be furnished with materials for the growth of frame and feathers; the laying hen for the making of eggs; the market fowl for the production of meat. The non-laying moulting hen requires foods rich in feather-making material. Young chicks eat more according to size than mature fowls. During the growing stage a large part of the food goes to produce frame or bone and feathers. The young fowl, or chicken, has less flesh or fat than the mature fowl. It has a smaller percentage of edible meat than the mature fowl. The reason is, the food is used more largely for frame building. The fowl, therefore, that is building a frame needs more frame material than one whose frame is already built. The skeleton of the fowl is made up largely of mineral matter—lime, phosphorus, iron, etc.—which are all grouped together under the name of ash in the ordinary food analysis.

The food requirement varies, therefore, as the stage of maturity varies. The young growing fowl requires more ash or mineral matter than the mature fowl that is not laying. The importance of this fact is brought home to the poultryman who keeps his fowls in enclosures where all their requirements must be met from the supplied food.

After the chicks have passed the brooding stage, which is usually at the age of six weeks to two months, depending upon weather conditions, they are past the critical period of their growth. If they reach this stage in good health and vigor, only mistaken feeding and management will result in stunted growth.

Management of Growing Stock.—While chicks may be successfully grown under more or less restricted conditions, the best practice is to give them free range. There are two reasons for this. First, the chickens are able to find feed that is often lacking in the ration when kept in confinement; and second, in hunting for feed they get exercise that they often do not get for lack of incentive when kept in yards. The importance of exercise for the growing stock cannot be over estimated.

Clean Range.—The range should be clean. It should not be overstocked with chickens. Where large numbers are kept the best conditions are obtained where fowls are kept in limited numbers in colony houses, separated widely so that they have plenty of clean ground to range over. There are many advantages of free range for growing chickens, among which may be mentioned: Less danger from contagious diseases; greater vigor due to greater exercise; greater profit because much feed otherwise wasted is found in the field; and the destruction of insects, such as grasshoppers, which may be an important item in certain localities. The chickens may run in the orchard, in the pasture fields with the cows and in the stubble fields after the crops are harvested. A flock of five hundred cockerels were kept on thirty acres of wheat stubble by the Oregon Station for two months in the fall, without additional food, the houses being moved several times.

Clean Yards.—Where it is necessary to keep the growing chickens in yards it is important that they be kept clean. If possible a crop should be grown on the yards every year. By plowing them and seeding them in the fall there will be a green crop in the spring on which the chickens may run. This will also help to keep the ground in a sanitary condition. If the yard is small, frequent spading or cultivation will lessen the danger of soil contamination and the fowls will scratch in the loose soil and get exercise in that way.

Shade.—Another essential of success in growing chickens is that they have an abundance of shade. Fruit trees or other trees may be planted in the yards, or part of the yards may be planted to corn or sunflowers. The latter make an excellent shade and at the same time furnish considerable feed. Where the shade cannot be secured in this way, artificial shade of some kind should be provided, such as frames covered with burlap or building paper.

Houses.—Ventilation or fresh air should be the first consideration in housing growing chickens. For a small house one side should be entirely open. If used in the cold weather of spring it would be an advantage to have the opening adjusted so as to prevent chilling during the cold nights when the chicks are small and not feathered fully. A house 7 x 10 feet will accommodate one hundred growing chicks or 5 x 8-foot house accommodate fifty chicks. Before they approach maturity the number should be reduced. The perches should be about 12 inches apart.

Size of Flock.—Where kept in colony houses on range, one hundred chicks in a flock should be about the maximum.

A house large enough to accommodate this number may be pulled by a team of horses. Keeping them in smaller flocks than fifty will offer no special advantage and the extra amount of labor in caring for them in smaller numbers offsets any possible advantage. The cost of the house will be greater in proportion to number for the smaller house than for the larger. The tendency to be guarded against is crowding too many together in a small house or coop. Great losses are incurred each year from this cause. The size of flock for the two houses mentioned may be 25% greater at the start and in two or three months, as the cockerels are marketed, the number will be reduced to the proper size.

Rations.—The feeding of the chick up to the end of the brooding stage has already been discussed. No sudden change should be made in the ration from small chicks to growing chicks. One of the great secrets in feeding chickens for any purpose is to avoid radical or sudden changes. Free-range chicks may safely be hopper-fed. Where the range is good, hopper feeding or part hopper feeding will give probably as good results as any other. A satisfactory method is to keep before the chickens all the time a hopper of dry ground grains and a supply of animal food. And in the afternoon or evening give a feed of whole grain, wheat or cracked corn, or a mixture of both.

The following are suggestions for a hopper of dry mash:

No. 1. Bran 3 pounds.   Ground corn 1 pound
No. 2. Bran 3 "   Ground oats 1 pound
No. 3. Bran 3 "   Ground barley 1 pound
No. 4. Bran 3 "   Coarse middlings
                                               or shorts 1 pound

If the animal food is beef scrap add 10 pounds of it to 100 pounds of the mixture. There should also be a hopper or box of broken or granulated bone and another of grit.

If the range is good and they find many bugs and insects, 10 pounds of beef scrap in 100 pounds of the dry mash will be sufficient. especially as they grow older and range farther.

If milk is available, that should be substituted for the beef scrap, though the addition of a small quantity of beef scrap to the mash makes it more palatable, otherwise the fowls may refuse to eat as much of the mash as may be desirable. The milk may be either fresh skim milk or sour milk or buttermilk, whichever may be the most economical or convenient to feed; either sour milk or buttermilk being preferable.

Moist Mash.—A daily feed of moist mash in place of the dry mash will result in more rapid growth. The real values, however, of the two methods have not been very clearly demonstrated. The difference will depend somewhat upon the character of the range and the amount and kind of insect food available. Either method, however, will give good results and whether one or the other method is used may safely be left to the convenience of the feeder. It will be better in feeding pullets to follow the method that will be followed in feeding them as layers. Changing from one method to the other at the beginning of the laying season will interfere with the laying for some time and a loss result. If necessary to make a change, it should be made gradually. The moist mash should be fed in the morning, preferably, and enough fed to last them about an hour. It may be made of the same grains as the dry mash, mixed, if possible, with milk or buttermilk.

Culling.—If the pullets have come from good breeding stock and have been properly hatched and brooded little culling will be necessary, but it would hardly be possible to find a flock in which some culling will not be desirable. It has been seen how important a thing is vigor, and though the lack of vigor may not be always apparent in the young stock, it is always apparent in some and when found the poultryman should cull rigorously. Where there is a considerable percentage of culls and this persists under best methods of hatching and brooding, the breeding stock should be changed. A change of males may be all that is necessary.

Poultrymen who use artificial hatching and rearing should set a hen or two at the time incubators are set and with the same kind of eggs and compare the results with those in the incubator. This will give a check on the breeding stock, as well as on the incubator. If the chicks under both methods show equally good growth and low mortality the poultryman should be satisfied that both the breeding stock and the incubation are all right. If, however, the chicks show poor results both in mortality and rate of growth, the evidence would point to the breeding stock as lacking in vigor; but if the hen-hatched chicks show good vigor and the incubator chicks poor vigor, the trouble is in the incubator or brooder. To determine whether the trouble is in the incubator or brooder, some of the incubator chicks should be brooded by hens. If the chicks show good vigor brooded under hens and poor vigor in brooders, the fault is in the brooding, not in the incubation.

Under best conditions, however, some culling will be found to be necessary. If at the age of two months some chicks have failed to make growth the poultryman will be money in the pocket if Dr. Hatchet is given a job. They are taking up room, eating food that will bring no return and are more or less of a menace to the rest of the flock.

The Cockerels.—A mistake is often made in retaining the cockerels too long. If they are hatched early in the season they will come to broiler maturity at a time when prices are at their highest for broilers. That is the time to sell them, usually. A broiler weighing a pound and a half will often bring as much money in April, May or June as a three or four pound cockerel in the fall. As weight is being put on the cockerel during the summer the price is falling, and the price often falls faster than the weight increases. By keeping the cockerels till the fall, therefore, or until they get their growth, the farmer or poultryman will very often get nothing for the feed he has fed them.

There are, of course, exceptions to this rule. A farmer may have a bunch of cockerels on free range where the food costs little or nothing. Where stubble fields are available till late in the fall and in sections where chickens can range out on stubble fields till near Christmas, it may pay to keep the cockerels over or until the time the prices have risen. Again, it may be that a farmer can caponize his cockerels in the summer and by keeping them till January or February, sell them at a good price for roasters.

Feeding Broilers.—If the chickens have been hatched early and it is desired to market the cockerels when the broiler market is good, the cockerels should be separated from the pullets when they weigh about a pound and given special feeding. In place of feeding the ground grain dry, as may be done with the growing stock, it should be mixed with sour milk or buttermilk. When feeding for flesh rather than for growth the proportion of bran should be reduced. For fattening broilers equal parts of bran and ground grain should be used, reducing the proportion of bran during the last week to one-half part. As much of the mixture as they will clean up in an hour should be given early in the morning. In the afternoon or evening whole grain should be fed. In other respects the same feeds should be given as for growing stock if the broilers have free range. If crate-fattened, they should be given soft feed exclusively for eight or ten days before marketing, feeding three times a day. Enough milk should be used to make the mash into a thin gruel and no water given to drink. No beef scrap or green feed will then be needed.

FATTENING OR FLESHENING FOWLS

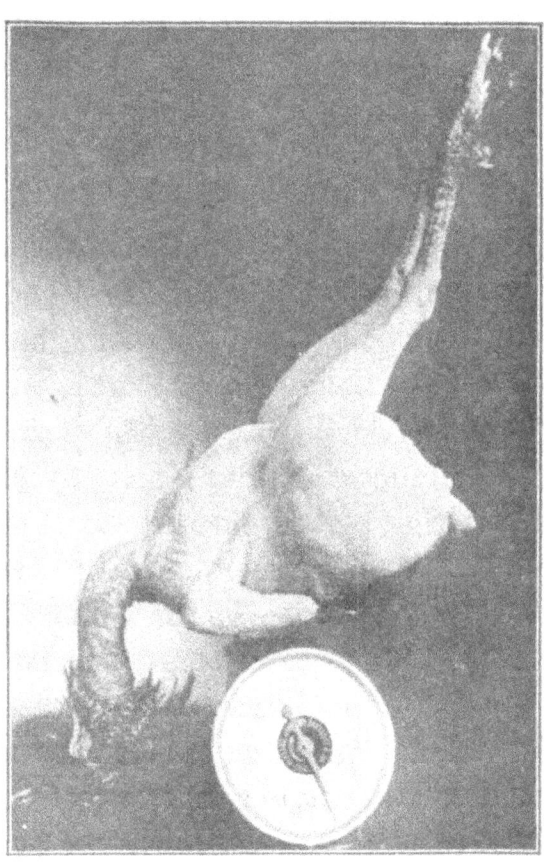

A GOOD ROASTER
8 1-3 months old, weighing 10¾ lbs. Fed on farm of Geo. H. Hyslop, Deslar, Ohio.

Special feeding before marketing greatly improves the quality of poultry. Unfattened poultry respond very readily to feeding. The period of fattening is about two weeks. In the case of beef animals it requires months of feeding to put them in condition for the best markets. The same evolution is accomplished for the fowl in two weeks.

In spite of the ease and rapidity with which the finishing or fleshening process is done, the great bulk of the poultry that goes to market lacks this finish. In recent years the fattening of farm poultry has been undertaken by the meat packers. The fowls are collected in large numbers and sent to feeding stations where, under proper conditions, great improvement is made in the quality of the chickens.

The objects of fattening may be stated to be, first, to add additional weight, and, second, to improve the quality of the flesh. The profit in feeding comes as much from the improved quality of the meat as from the additional flesh put on. Fattening is especially desirable for young cockerels that have had free range on the farm. They have good frame and constitution, and when confined and properly fed put on flesh rapidly and economically. The farmer might well secure the benefit of the extra weight and the consumer the extra quality. The skillful feeder feeds for both quality and weight. The fowl that is simply fat has the fat distributed over the intestines and under the skin and when cooked this fat will run out into the pan. With the fowl properly fattened the fat will be distributed in small globules throughout the fibres of the flesh and when cooked the flavor of the meat will be retained and the meat will be more tender. The consumers in purchasing fowls at so much per pound are paying for bones as well as meat, and they prefer the fattened fowl at a higher price because they get more edible meat in each pound purchased. Proper fattening increases the proportion of meat to bones, and this is the special benefit of fattening.

Methods of Fattening.—There are three methods followed in fattening. First, pen fattening; the fowls are confined in small pens or yards. Second, crate fattening; by this method specially made crates or feeding batteries are used. Third, cramming; in the last stage of the fattening period a cramming machine is used.

The first method is largely used on the farm where the business does not receive special attention. The second method is used at the large packing-house stations and by others making a special business of marketing fowls of extra quality. The cramming method is not very generally used in this country. In England and France great numbers of chickens are "crammed."

In egg production one of the essential factors is exercise

for the hen, while non-exercise is just as essential in fattening. The good laying hen must have vigor and this is associated with hard muscles. Proper fattening means a softening of the muscles to produce a flesh that when cooked is tender, and this is produced only by restricting or preventing exercise. In fattening it is not the object to secure vigor in the chicken; rather it is the initial process leading to loss of vigor. The fattening process could not continue long beyond the two weeks fattening period without the fowl showing decided loss in vigor.

FEEDING BATTERY FOR FATTENING

Another essential in fattening is that soft foods must be fed altogether. Lack of exercise interferes with the proper digestion of whole or hard grains. The grain is ground finely and mixed with water or milk to about the consistency of cream, or thin enough so that it will drip from a spoon. No water is given to drink.

Cost of Fattening.*—The cost of feed consumed by 498,681 chickens at four large packing-houses in the middle West in 1912 was as follows:

---

*Bulletin 21, U. S. Department of Agriculture.

	Cost of feed per pound gain	Cost of labor per pound gain
Packing house A	8.74 cents	1.63 cents
Packing house B	7.70 cents	1.99 cents
Packing house C	6.61 cents	1.37 cents
Packing house D	9.95 cents	1.59 cents

The cost was figured on the following prices of feed:

Corn meal	$1.39 to $1.74	per 100 pounds
Low grade wheat flour	1.38 to 1.52	" "
Shorts	1.18 to 1.25	" "
Buttermilk	1 to 2 cents a gallon	

Fattening Rations.—The best results were secured by feeding either of the three following rations:

1. 3 parts cornmeal. 2 parts low grade wheat flour; 1 part shorts.
2. 3 parts cornmeal, 2 parts low grade wheat flour.
3. 5 parts cornmeal, 3 parts low grade wheat flour, 1 part shorts, and 5 per cent tallow.

Oatmeal produced better gains than low grade wheat flour, but was less profitable on account of its higher price.

Buttermilk is used by the packing-houses in mixing the food. No other animal food is given. Buttermilk or sour milk is preferred to sweet milk. The milk or buttermilk bleaches or whitens the flesh. All milk-fed chickens have light-colored flesh. This whitening may be partly offset by feeding yellow corn meal. If milk is not used, beef scrap or other animal food must be fed.

The feeder must use the foods that are reasonable in price. No one grain is essential. In most sections of the United States corn will be most largely fed because of its cheapness. In the fattening districts of England oats are considered the most satisfactory. In France buckwheat and barley are largely used. Where oats that are good and

FEEDING STATION

Here chickens are fed ground grains and buttermilk.
(Courtesy Bureau of Chemistry, U. S. Department of Agriculture.)

plump and cheaper than other grains can be secured, they should form the large part of the ration. They must, however, be specially ground to cut up the hull in small particles. If corn and oats cost the same pound for pound, then the ration may be made up of half of each by weight. If corn is not available, a little middlings or shorts, or low-grade wheat flour, may be mixed with the oats. The famous Sussex fat chickens in England are produced on a ration of oats. The important thing, however, is that the oats be heavy and finely ground.

CHAPTER XIV

METHODS OF HATCHING CHICKENS

Structure of the Egg.—The principal parts of the egg are, in the order of their growth, the yolk, the albumin or white, and the shell. The yolk is built up carefully layer upon layer and requires about two weeks to develop from the size of a pea to the full-sized yolk. It contains the blastoderm or germ cell, which may be seen as a white speck one-eighth inch in diameter on its upper surface. This speck enlarges when the egg is kept in a warm room or in a high temperature. The blastoderm, as Lillie says, "is the living part of the egg from which the chick embryo and all its parts are derived." There is more or less development of the embryo of a fertile egg before it is laid, due to the body temperature of the hen. Should the egg be retained in the uterus, as sometimes happens, a day or two before being laid the development may proceed so far that the egg will be unfit for eating. Retarded laying, however, seldom happens. The yolk furnishes the embryo a large part of its nutriment, and the unassimilated part of the yolk furnishes the chick food for several days after hatching.

The Albumin in different layers surrounds the yolk. Close to the yolk there is a dense layer which forms at each end of the egg two spirally twisted cords. These are called the chalazae, the apparent function of which is to hold the yolk in place. The albumin is a protection for the germ or blastoderm. It keeps it from coming in contact with the shell and lessens the force or effect of jarring. Another function of the albumin is to prevent the entrance of bac-

282 POULTRY BREEDING AND MANAGEMENT

teria to the yolk or germ cell. The oviduct appears to be germ-proof and the albumin to have certain bactericidal properties.

The Yolk has a lower specific gravity than the albumin and will be found floating near the side of the egg uppermost, with the germ cell on the upper side of the yolk. This brings the germ cell always near the source of heat during incubation. Should the egg remain long in one

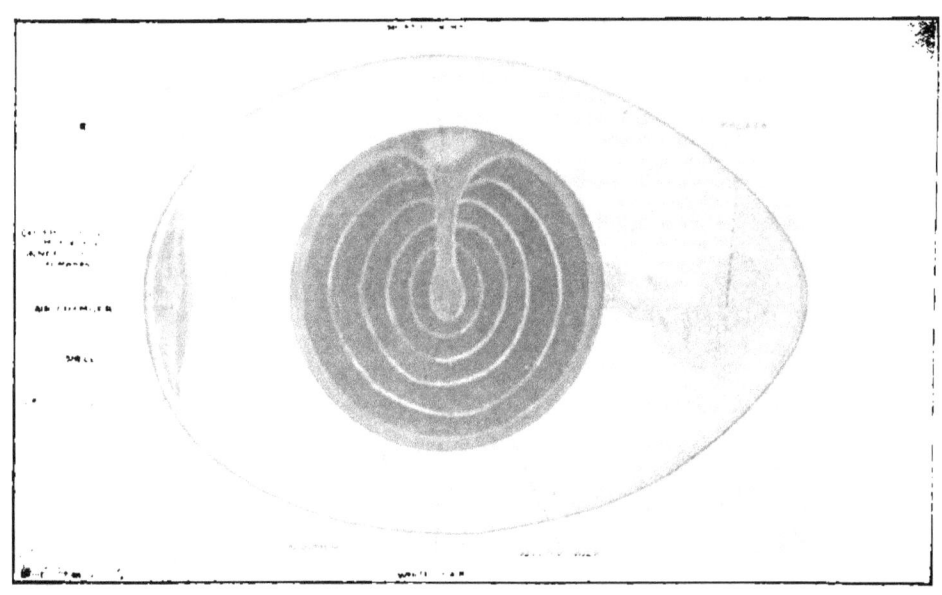

PARTS OF A FRESH EGG

position without turning, the albumin becomes thinner and the yolk will adhere to the shell. This is fatal to the embryo.

There are two layers of *shell membrane,* the inner and outer. The inner layer lies next to the albumin and the thicker outer one next to the shell. After being laid, the egg contents contract, and at the large end the inner layer draws away from the outer, causing the air space.

The Shell, which forms about 11% of the weight of the egg, plays an important part in the embryonic development of the chick. Until recently the shell was supposed to be merely a protective device, but it is now known that the developing chick draws upon the shell for a large part of the lime necessary for its proper growth. The shell is porous or permeable to gases; this permits evaporation of the water of the egg and also permits entrance of oxygen necessary for the development of the embryo.

How Long Should Laying Hens Be Kept.—The productive life of the hen is short compared with that of domestic animals. It is apparently a natural characteristic of the hen to lay more eggs in her first laying year than in the second, and in each succeeding year a smaller number. The results of experiments by the writer at the Utah Station showed average pen results from Leghorns in the first year of 164 eggs, and in the second 126 per hen. In exceptional individual cases more eggs were laid the second year than the first. For instance, one hen laid 201 the first year and 241 the second. Forty-one hens of different breeds averaged 178 eggs the first year and 125 the second, or 40% more the first than the second year.

At the Oregon Station later results were secured as follows: Fifty Barred Plymouth Rocks laid 160 eggs the first year, and 105 the second year, and 50 White Leghorns and crosses, 153 eggs the first year, and 130 the second year. In these experiments the laying year began November 1. Out of the 100 hens, 17 laid more eggs the second year than the first.

Other records showed, what might be expected, that where the conditions for egg production were more favorable during the second year than the first, a better egg yield was secured during the second year. Again, where the period of maturity varied, or where the laying year began in the spring, the second year records were better than the first. The moulting period did not vary, making the first laying year short

THE BEGINNING OF THE END OF INCUBATION
The prison walls being broken down.

Where maturity is reached in the fall and laying begins then, the average flock results invariably showed that the first year is the most productive or profitable, and that there is a gradual decrease each succeeding year.

This means that the hens must be killed off at the end of their second laying year and their places taken by pullets. Some poultry-keepers practice renewing the flock of layers every year; others keep them three years.

Renewal of the Flock.—The point is that the frequent renewal of the flock constitutes a large and costly part of the business of egg production. If the productive life of the hen could be lengthened to say five years instead of

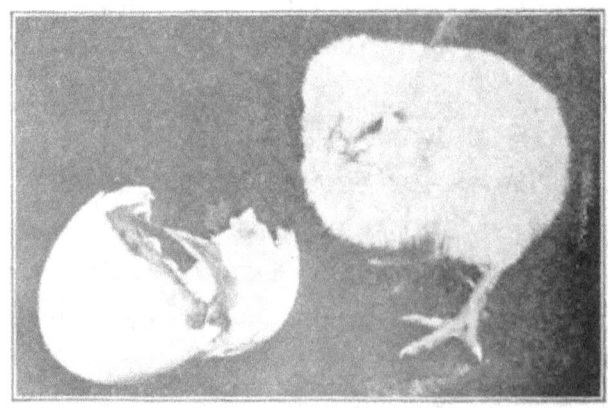

THE YOUNG GRADUATE
It has broken its way out into the world—the fruits of successful incubation.

two years, one of the most troublesome as well as expensive features of poultry-raising would be very much simplified. The cost of incubation, or the hatching and raising of the pullet, is a no small initial charge on the cost of every dozen of eggs produced. This initial cost would be less significant were it possible to eliminate a large part of the losses usually incurred in the hatching and rearing of the chicks.

In renewing the flock the utmost care must be taken in hatching and rearing to preserve in the new flock the vitality of the old. If the health and vitality of the stock may be injured by improper methods of incubation and brooding, and these methods are persisted in year after year, disastrous results will soon be brought about by the very frequency with which the flock is renewed. Decrease in egg production, which we may seek to overcome by frequently renewing the flock, will as certainly result from a gradual lowering of vitality as from keeping the hens till they have "lost their teeth." It is the opinion of the writer that there are harder problems to solve, and greater difficulty to be encountered by the poultryman, in incubation and brooding, than in any other part of the poultry business.

Eggs for Hatching Must Be Produced by Hens of Good Vitality.—Successful rearing of chickens depends very largely on following closely nature's way. If we study the way of the hen that hatches every egg in the fence corner, we shall find this fact: The hen that laid the eggs was not confined in close yards; she had the liberty of the fields. This guaranteed good health and vigor. Eggs laid by such hens will hatch better than those from hens cooped up under artificial conditions. Health and vitality in the hen are transmitted to the chick. Eggs that hatch well come from hens that have good vitality. Chicks that live well come from eggs laid by hens of good vitality. The method of hatching or the method of brooding is not always responsible for eggs failing to hatch and for chicks failing to live or grow well. The parent stock, or the condition under which the parent stock is kept, is sometimes to blame.

Breeding stock, therefore, should be carefully selected, only those individuals being retained that are up to a certain standard of shape, size and vigor.

It is not claimed here that lack of vigor in the parents will inevitably be transmitted to the offspring. Parents of apparently weak constitution may breed vigorous offspring. A chicken may have been injured in its rearing and show weakness, without, however, impairing its value as a breeder of strong, healthy stock; but the poultryman cannot afford to retain in his flock fowls showing constitutional weakness. Lack of vigor in the parent stock may not always show in the offspring, but it will invariably show itself in smaller egg production and in eggs that do not hatch a high percentage of chicks. There may not be constitutional weakness in the fowls that lay the eggs, but if there is lack of vigor there will be correspondingly few eggs that are fertile and fewer of the fertile eggs that hatch.

Methods of Hatching Sometimes Responsible for Poor Hatches and for Lack of Vigor in the Chicks.—Do not always blame the parent stock for poor hatches and for poor chicks. At the Oregon Station one method of hatching gave an average of 78.8 chicks from a hundred eggs set, while another method gave 60.6 chicks. When brooded in artificial brooders, 90% of the chicks hatched by the first method were alive at the end of four weeks, while only 67% of the others were alive. When brooded under hens, about 98% of those hatched by the first method were alive at the end of four weeks, and only 51% of the others.

While it is true, therefore, that lack of vigor in the

parent stock may sometimes account for poor hatches and low vitality in the chicks that hatch, it is also true that poor methods of incubation may produce the same results.

Methods of Brooding are sometimes faulty and result in a high death-rate among the chicks and in impaired vitality in those that grow to maturity.

Feeding and General Care of the chicks is an important part of this subject. It is true that chicks of good vitality will stand a good deal of abuse in the rearing; it is true that expensive foods and much labor in feeding are not necessary to get the best results; but at the same time, to get the rapid growth required of the chicks, they must have proper foods.

With the above outline as a guide, let us now discuss some of these points more in detail. Omitting further reference to the first topic, let us consider different methods of hatching.

NATURAL *VERSUS* ARTIFICIAL INCUBATION

There are two ways or methods of hatching chickens, namely, natural and artificial; in other words, hen-hatching and incubator-hatching. On the general farms the larger part of the hatching and brooding is done by hens, while a majority of the special poultry farms use incubators and brooders.

There are advantages and disadvantages in each. The poultry-raiser must choose the method that best suits his individual conditions. Each method has its place, but there is a difference of opinion as to how far artificial methods should supersede the natural. On the general farm where fifty or a hundred fowls are kept the natural method is undoubtedly the most satisfactory, first because of the limited number of chicks to be hatched to renew the flock; second, because farmers, as a rule, have not the time to give the necessary care to the incubator and brooder; third, because the cost of equipment is much less; and fourth, that under farm conditions better chicks will be reared by the natural way.

The advantages of the artificial method are mainly apparent on the large special farms. Incubators are a necessity on these farms first, because non-sitting breeds are kept on many of them; second, because not enough sitters can be secured early in the season for hatching chicks to supply the market with early spring broilers; third, it is claimed that the incubator lessens the labor where large numbers are hatched. A fourth, and important advantage, is that the use of incubators makes it possible to keep the chicks free from lice and mites and certain diseases.

There are, however, large poultry farms, where egg production is the chief object, that use the natural method, noticeably that of the Little Compton district of Rhode Island. Artificial methods would be more generally used than they are were it not for the fact that there are problems in artificial incubation and brooding that are not encountered in natural incubation and brooding. Whatever may be the real merits of the two ways of hatching, it is certain that the incubator has become a considerable factor in the poultry industry, and it may be that with improvements in manufacture and methods of operating the machine it may in the future still further supersede the hen.

Comparisons of the Two Methods.—Reports of experiments on the relative efficiency of the hen and incubator, are somewhat contradictory. In tests at the Oregon Station that extended from April to July comparative results were secured. These experiments were made in the spring and summer months, and it has been the experience at this station that incubator chicks hatched earlier in the season, or in the colder months, have greater thrift. Chickens hatched by incubators in January, February and March are more easily reared than those hatched later. It is not assumed that the results secured were the best that may be obtained. It was an incubation experiment. Chicks hatched in different ways were put under like conditions of brooding, and even though the brooding might not have been the best, the value of the incubation comparison should not thereby be lessened. The following is a summary of the results:

1. From 879 eggs set, incubators hatched 533 chicks, or 60.6%.

2. From 279 eggs set, hens hatched 219 chicks, or 78.8%.

3. Eliminating eggs broken in nests, the hen hatched 88.2% of eggs set.

4. The incubators hatched 78.5% of "fertile" eggs, and the hens hatched 96.5%.

5. Eggs incubated artificially tested 22.7% as infertile, while those incubated by hens tested out 11.8%.

6. The incubators showed 16.6% of chicks "dead in the shell," and the hens 2.8%.

7. Chicks hatched under hens weighed heavier than chicks hatched in incubators.

8. The mortality of hen-hatched chicks brooded in brooders was 10.8% in four weeks, and of incubator-hatched chicks 33.5%.

9. The mortality of hen-hatched chicks brooded under hens was 2.2%, and of incubator chicks 49.2%.

10. In other tests the mortality was 46.5% for incubator chicks brooded by hens and 58.4% brooded in brooders.

11. Hen-hatched chicks made greater gain in weight than incubator chicks, whether brooded by hens or brooders.

At the Ontario Agricultural College, experiments gave the following results: "Nine hundred and fifty-eight eggs were set in the machines and 436 chicks were hatched, or 45.5% of the eggs set. Three hundred and thirty-five were set under hens and 196 chicks were hatched, or 58.5% of the eggs set. As the same hens' eggs were used in each method, the hen has the advantage, and had she not been in cramped quarters for a portion of the hatches, her hatches would have been larger." (Ontario Agricultural College, Bulletin 163.)

Prof. Edward Brown, President of the International Association of Poultry Instructors and Investigators, in the "London Illustrated Poultry Record," says: "The most ardent advocate of artificial methods of hatching cannot but acknowledge that there is something yet to learn, or rather that incubators are second best and hens are first."

Natural Incubation and Brooding.—By the natural methods of hatching, the cost of the incubator is eliminated, hens taking its place. In the same way brooders are dispensed with. Hens may be set and chicks reared in one coop, or house, such as is illustrated on page 293. Assuming that it requires as many eggs to hatch one hundred chicks with hens as with an incubator, three coops costing not more than $8 each will be required, each coop accommodating four sitting hens. By setting the hens at one time the chicks hatched in one coop may be given to two hens to brood. The hens need not be included in the cost of the equipment, for they will be worth practically as much after hatching and rearing their chicks as before. Neither is it necessary to charge the method with eggs that might have been laid by the hens if they had not been used for hatching. It is doubtful if the hen will not lay as many or more eggs during the year if she has hatched and reared a brood of chicks than if her natural instincts had not been gratified. By taking a month or six weeks off for hatching in the spring when eggs are cheap, the hen is usually in better condition and will begin to lay earlier in the fall when eggs are a good price. It is not certain, therefore, that the yearly production of a hen will be lessened by allowing her to hatch and rear a brood of chicks.

Two Methods of Hen Hatching.—If a proper system be followed, chickens may be conveniently and successfully

NESTS USED FOR SITTING HENS

Showing apparatus used for drawing air from under sitting hens for determinations of humidity and carbon dioxide in the nests. (Oregon Station.)

reared in large numbers by natural methods. In hatching with hens a system should be followed that will economize in the labor in caring for the hens. On most farms very little attention is given to furnishing convenient hatching facilities. A little thought and a few dollars spent for equipment will make the work less troublesome and uncertain. One of two methods may be followed. The first requires daily attention in letting the sitters off the nest and seeing that they go back again. By the second method the hen leaves the nest and returns at will.

Where it is desired to set a large number of hens, they will be conveniently looked after by making a bank of nests along the side of a poultry house or in some unused shed. The nests should be about 12 x 12 x 14 inches in size, made by taking two 12-inch boards for the top and bottom and cutting another 12-inch board into 14-inch lengths for the partitions, then nailing them together, as many as desired. The top of the bottom row will furnish the bottom of the second row, and four or five rows of nests may thus be placed together. There should be a hinged board at front to confine the sitters. The hens should be let out every day to eat and drink for about 15 minutes, the length of time depending on weather conditions. Several inches of fine waste hay should be placed in the bottom of the nests. Short-cut straw or clean chaff will answer the purpose.

The second method of hatching and brooding by hens requires less care on the part of the attendant. We have found it to work well at the Oregon Station. One coop serves for both hatching and rearing the chicks. It serves the triple purpose of an incubator, a brooder, and a colony house.

A convenient size of coop is 5 feet long and 3 feet wide, with a shed roof 3 feet high at front and 2 feet at back. It is large enough to divide into separate apartments for four sitting hens. Movable partitions of canvas or burlap are fastened to a 4-inch or 6-inch board at the bottom and to a cross-piece at the top. It has an outside run 3 feet long for each hen, covered with wire netting. The runs are hooked on to the house and may be dispensed with when the chicks are hatched. These runs give the hens opportunity for dusting and exercise. By keeping feed and water before them all the time, the sitters may be allowed

METHODS OF HATCHING CHICKENS

to leave the nests and return at will. In this way, very little labor is required in caring for sitting hens. The door on the front is hinged at the top. Underneath this there is a frame of wire netting of 1-inch mesh; in mild weather the door may be kept open. This makes it an open-front house, and during the summer months, when the growing chicks use it as a roosting-house, it will be found that this provision for fresh air is a necessary one. More conveniently to get at the nests or sitters, there is an opening at the back of the coop. The top board, which is

(1) Arranged for hatching, with doors closed.

(2) Showing door and top open.

(3) Arranged for brooding, with runs detached.

THE OREGON STATION COMBINATION HATCHING AND BROODING COOP

For four sitting hens, each in separate apartment.

10 or 12 inches wide, is hinged at the bottom and cut in the center, one door serving for two nests. It is also an advantage to put hinges on the roof so that it may be opened. In dry locations and where the house can be moved to fresh ground occasionally, floors are not necessary or desirable. In some localities where rats are a pest, a floor will serve to prevent losses of eggs and chicks.

When the chicks hatch, about twenty of them may be given to a hen to brood, and the remaining hens reset. The hen and chicks may be put in a coop such as is shown on this page. This will afford them shelter for a month or six weeks, after which a larger coop in which they can be brought to maturity will have to be provided. Or the small coop may be dispensed with, and the hen and chicks transferred at once to the larger coop. A coop the size of that illustrated on page 293 is large enough to rear the chicks of two hens, or forty chicks.

A NEW BROOD COOP
Oregon Station.

A New Brood Coop.—A brood coop which may also be used for sitting hens is shown on pages 294 and 295. It is ratproof and rainproof. The wire front affords provision for fresh air. The netting is of 1-inch mesh. The door which slides up and down may be fastened with a wooden pin, to allow the chicks to come out and confine the hen. The bottom is separate to afford easy cleaning. To clean, the floor is pulled from under the coop, or the coop lifted to one side. Where there is no danger from rats or

METHODS OF HATCHING CHICKENS

other vermin and the ground is not too damp, the floor may be dispensed with. The roof is likewise separate. This makes it more convenient to catch the hen or the chicks. The roof is made of flooring and the sides of shiplap. To prevent rain beating into the coop and also to protect from the sun in hot weather, a shield is provided, hanging at different angles. When not in use this shield is pushed back under the roof. To keep the shield from falling down when it is pulled out, two pieces of strap iron, with proper bent, are fastened to the board. For further details, see working plan, this page.

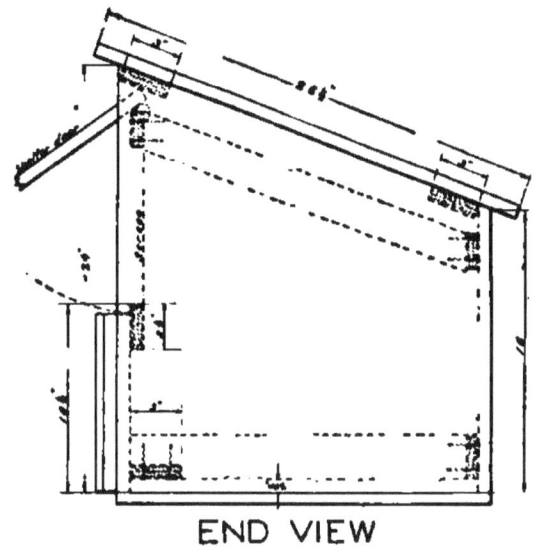

END VIEW

PLAN OF HEN BROOD COOP

Points on Setting a Hen.—The best sitters are the breeds of the American and Asiatic classes. The Mediterraneans, such as Leghorn, Minorca, Andalusian, are not good sitters. Hens of gentle disposition should be chosen if possible. The normal temperature of the hen's body is about 106 degrees. One good broody hen will take care of twenty chicks. Another advantage of setting several hens at a time is that the chicks may be "doubled up." For an average hen, 13 eggs are enough for a sitting.

Dust the hen with insect powder when taking her off the nest. Examine the heads of the chicks two or three days after hatching, and if lice are found, rub a little lard on the head and under the throat. If the hen has been properly treated for lice while sitting, there will be no necessity for treating the chicks for lice. But watch them.

Moistening the eggs before hatching is not necessary.

HEN BROODING AT OREGON STATION

The hen attends to that herself. Keep the hen and chicks on clean grass runs if possible. If properly managed, the hen may be got to laying after being with the chicks two or three weeks. In warm weather the chicks may be "weaned" when a month old.

Feed the sitters corn or wheat, all they will eat, and provide grit, water, and a little green food. Provide also a box of earth for dusting; earth should not be too dry.

The hen will usually hatch best in a nest on the ground, but the ground should not be too hard. Cover it with chaff, or short straw or hay. If the hen is set on a board floor, put in two or three inches of moist earth, hollow the nest slightly in the center and cover with straw or hay. Planer shavings are also good nesting material.

Dust the hens with a good insect powder or tobacco dust when setting them, and again ten days later rub it well into the roots of the feathers. Put a spoonful of the same material, or a moth-ball, in the center of the nest. With this treatment the hen should be free from these pests during the period of incubation, if the house or box in which she is sitting is not infested with them. If it is found necessary, dust the hen oftener.

ANOTHER TYPE OF BROOD COOP
Made out of a shoe box.

Period of Incubation.—The period of incubation for different species of poultry is shown in the following table:

Species	Days	Species	Days
Hen	21	Peafowl	28
Pheasant	22-24	Guinea	26-28
Duck	28	Ostrich	42
Duck (Muscovy)	33-35	Goose	30-34
Turkey	28		

Selecting Eggs for Hatching.—It is not possible to determine whether a fresh egg is fertile or will hatch. The egg must be under the sitting hen or in the incubator several days before its fertility may be determined. Neither is it possible to tell from any differences in shape of the egg whether it will hatch a male or female chick. The shape or size of the egg has nothing to do with the sex of the chick. There are, however, certain points in shape and structure of the egg that should be considered in selecting eggs for hatching. Normal eggs should be selected. This does not mean that the eggs should all be of the same size. Eggs laid by different hens vary in size even when

BROODING COOPS

The brooding coops on the farm of F. W. C. Almy, Rhode Island, where 5,000 chicks were hatched and brooded by hens in a year.

the hens are of the same breed. One hen may lay an egg weighing more than two ounces; another, less than two ounces.

The most profitable hen is not necessarily the one that lays the largest egg. The hen that lays a small egg may produce so many more of them in a year that she will lay a greater weight of eggs in a year even though her eggs average much less in weight. The large egg may be normal for one hen and the smaller egg for the other. Other things being equal the one will hatch as well as the other. The size of egg is a matter of breeding or heredity. It is well to use the larger eggs for hatching, because in that way it will be possible in a few years to breed up a strain of fowls that will lay larger eggs. Abnormally large or small eggs should not be used for hatching. Eggs that are not normal in shape should also be discarded. Ill-shaped, rough-shelled, dirty eggs should not be used.

It is very important to select fresh eggs, the fresher the better. It is possible to keep eggs several weeks and have them hatch, but eggs lose in hatching quality the longer they are kept. They will keep in a cool place better than in a warm place. They should not be kept in a moist, damp room. It is a good plan to turn them once a day and to handle them with clean hands.

There is great difference in eggs in fertility and hatchability. One of the chief causes of infertility in eggs is close confinement of the layers. Experiments have shown that eggs produced by fowls on free range are more fertile and hatch better than those from fowls confined in yards. (West Virginia Bulletin 71.) In these experiments about three times as many eggs tested infertile from the confined fowls as from those having unrestricted range. Whether the increased fertility from the latter was due to possibly greater exercise or to natural foods found on the range, the experiment does not show. So much importance, however, is placed on this point that many of the large hatcheries refuse to use eggs that have not been laid by hens that enjoy free range.

Testing the Eggs.—After six or seven days of incubation, the infertile eggs may be taken out and saved for the chicks. Rotten eggs should also be removed. If a number of hens are set at one time, they may be doubled up after testing; that is, if as many as a sitting are tested out, one hen may be reset on fresh eggs. An egg-tester may be purchased from a poultry supply house, or a small box may be used in which an electric light globe may be put. A coal oil lamp or candle may be used instead of an electric light bulb. (See pages 353-354.)

In one side of the box cut a hole about the size of an egg. Testing is done in a dark room or at night by holding the egg to the light at the hole in box. Eggs may be tested by daylight by holding the egg at the end of a tube or funnel and pointing it toward the light. Where the incubator room is fairly dark the eggs may be held at a hole in the wall and tested. With two holes a little smaller than an egg, two eggs may be held up at one time.

In testing incubator eggs the tray may be taken outdoors in the light and by using the funnel looking down on the eggs they may be quickly and easily tested. Another method of testing without handling the eggs is to use a flashlight under the tray in a dark or moderately dark room and looking down on the eggs direct-

DIVISION OF POULTRY LABOR

Some of the poultry farmers at Petaluma, Cal., take the eggs to the hatchery and pay the hatcher 3 or 4 cents apiece for hatching the chicks. The photo shows part of a hatch of 1,500 chicks at a Petaluma hatchery, which are to be taken by another party to raise.

ly above the flashlight. The latter two methods obviate touching the eggs with hands. An infertile egg will look clear, just like a fresh egg, only it has a little larger air cell. A fertile egg will show dark.

Artificial Incubation.—The hatching and rearing of chicks by artificial means has been practiced by Egyptians and Chinese for centuries. The ancient methods were crude, and their success depended upon skill obtained by long years of practice. The secret was handed down from

THE SAME CHICKS LOADED ONTO THE WAGON OF THE MAN WHO CONTRACTS TO RAISE THEM

father to son. No thermometer was used, the temperature being judged by the "feel" of the operator. Large hatching "ovens" were used. The eggs were purchased, and the chicks sold for about $1 a hundred. The business of hatching was confined to a few hatcheries or a few families who appeared to have a monopoly of the business. The same methods are followed to-day in Egypt, China and other countries.

In Europe and America artificial incubation is of comparatively recent origin. So far as a practical application is concerned its history goes back less than 50 years. Incubators had been used before. The first of which there is account was that of Reaumur, a Frenchman, who made and used a machine in 1749. Development has been along different lines than in Egypt and China. The monopolistic tendency is absent; the effort is to place machines on the market whose essential points are ease of operation and availability to all poultry-raisers. In recent years, however, there is a distinct tendency toward centralized hatching where the business is turned over to large hatcheries, and the poultrymen purchase the chicks as they come from

BROODER HOUSE

The chicks were loaded onto a wagon by the man who contracts to raise them, taken two miles into the country and put in brooder houses shown above. The farmer takes the eggs to the hatchery, where Mr. A., the hatcher, hatches the chicks at so much per, and Mr. B. takes them and rears them at so much per.

the incubator. In place of incubators with capacities from a few dozen eggs to two and three hundred eggs, there are now machines with capacities of 3,000 to 10,000 eggs.

The Incubator House.—For successful hatching the temperature of the room in which the incubator is operated must be controlled within certain limits. The first requirement of an incubator room is a fairly even temperature. The less the temperature varies, the more easily may the temperature of the incubator be maintained. It is not necessary that the room have the same temperature night and day. Within a reasonable range of temperature, say 10 degrees, the temperature inside the incubators as now constructed may be fairly well maintained.

Ventilation of the Room.—Of equal importance is the ventilation of the room of the incubator house. The developing chicks make a constant demand for oxygen, and

OREGON STATION INCUBATOR HOUSE

unless the ventilation of the room be ample the chicks will not develop and hatch with high vitality. It is the opinion of the writer that low vitality in the chicks is often the result of an insufficient supply of oxygen during incubation, more often than the result of variations in the temperature of the incubator house or in the machine, though that also is important.

A cellar or underground room is often used for the incubator. While it may afford the best conditions for controlling the temperature, it offers the poorest conditions for maintaining proper air purity. Fresh air is as important as a uniform temperature, and success will not be made where one or the other of these factors is absent. During cold weather there will be less difficulty in securing good ventilation, because there is a more rapid exchange of air where the difference between the inside and outside temperature is great. In warm weather when the temperature inside and outside the incubator room is about the same, the air will be stagnant, and the growing embryo will suffer from a lack of oxygen.

Analysis of air in an underground incubator cellar at the

INTERIOR OF OREGON STATION INCUBATOR HOUSE
On the left tests are being made of carbon dioxide in the incubators.

Utah Station while incubators were running showed as high as thirty parts carbonic acid gas in 10,000 parts. This is undoubtedly much beyond the limit of safety. At the Oregon Station in an incubator room above ground, analysis showed 9.9 parts, the highest, and an average of 7.5. Tests in May showed more carbon dioxide than in April.

This is a possible explanation of the fact that chicks hatched in the cool weather have better vitality than those in the warm weather of summer. Incubators are now built with sufficiently sensitive temperature regulation that it is not essential that the temperature of the room be so constant that an underground cellar is necessary. A house above ground, properly constructed, is more desirable than an underground room or cellar. In warm sections of the country it will aid in maintaining a more uniform temperature if the house can be shaded by other buildings or trees.

Choice of an Incubator.—There are probably half a hundred different incubators in general use, and many of them differ only in name. The essential features of the

A 150-EGG INCUBATOR, KEROSENE LAMP HEATED

majority of incubators are alike. The source of heat is usually a kerosene lamp. Others are heated by gas or electricity. While the source of heat may be the same, the methods of distributing the heat over the eggs varies in different machines. Because of this difference, incubators have been divided into two classes; the one in which the heat is radiated from a tank of hot water over the eggs, or from hot water pipes; the other type, in which, the hot air is diffused through cloth or muslin directly over the eggs. In the former, the hot water tank heats air already in the egg chamber; in the latter, fresh air is heated by the lamp as it enters the machine and the same air enters the egg chamber by diffusion. So that there is the diffusion type of machine and the radiated heat type. In the latter the ventilation is independent of the source of heat. It

A 250-EGG INCUBATOR, KEROSENE LAMP HEATED

requires longer to heat up the hot water machine than the hot air.

On the other hand, the former maintains the heat longer, and this is an advantage in case the lamp goes out. To secure the same ventilation, or the same rate of exchange of air in the hot water machine as in the hot air, or the same degree of air purity, there must be more openings for ventilation in the former than in the latter. If the ventilation be arranged, therefore, so that there may be the same rate of exchange of air, or the same degree of air purity, there should be little, if any, difference in efficiency due to

METHODS OF HATCHING CHICKENS

the different methods of furnishing heat to the eggs in the two types of machines.

The tank in the hot water machine adds to its cost. The tank is also likely to get out of repair and leak after a few seasons' use. The hot air type of machine is more generally used, and on the whole is the more satisfactory. Whatever may be the type of machine decided upon, however, one that is well made should be chosen.

Size of Machine.—Machines vary in capacity from fifty eggs to several thousand. Those of less than 100-egg capacity have not proven as satisfactory as larger machines. A machine of 125 to 150 capacity should give as good results as larger machines. A machine of this capacity may be used for any number of eggs less than that. It is a question whether the machine would not hatch better if not crowded to its full capacity. The size of machine to select should be governed, first, by the number of chicks it is desired to hatch, and, second, by the number of suitable fresh eggs that may be secured. Where chicks are to be hatched for fall and winter layers they should all be hatched at about the same time, or within a period that will bring the pullets to maturity at the proper time for laying. On this account it will be better to confine the hatching to two runs of the incubator. That would make a difference in about four weeks in the ages of the two lots of pullets. A mistake is made in hatching too early as well as in hatching too late where good fall and winter layers are desired. If the purpose is to produce about one hundred pullets for fall and winter laying, two runs of a 150-egg machine will be necessary. As to the egg supply, the fresher the eggs the better, and an assured supply of eggs of good hatching quality is necessary before purchasing the incubator. (See page 299.)

Operating the Incubator.—Space need not be taken here with elaborate directions for running the machine.

The purchaser of an incubator is furnished directions for its care, and those directions should be carefully followed until, at least, experience has demonstrated that they may be modified with advantage. Second, the construction of machines varies more or less, and no set of directions will suit all machines. The safe plan, therefore, is to study the directions that come with the machine. While different machines require different methods of operating, on some fundamental points, however, there should be agreement in directions.

Humidity Conditions.—On the question as to whether moisture should be supplied to the incubator, there is a great diversity of views among incubator makers. In experiments by the writer at the Utah and Oregon Stations, it has been found that moisture or humidity conditions have a great deal to do in the successful incubation of hen eggs.

Experiments reported in Utah Bulletin 92 (1905) showed that there was a greater loss in weight of eggs in incubators than under hens during incubation. This loss is largely water evaporated. In the first 18 days of incubation the average loss in incubators was 18.4%, and of eggs under hens the loss was 15%. In later experiments at the same station (Bulletin 102, 1907) machines with no moisture averaged 17.8% loss, medium amount of moisture 14%, and with maximum amount of moisture the loss was 12.3%. At the Oregon Station, as reported in Bulletin 100 (1908), eggs under sitting hens in dry nests averaged 14.8% loss. In later tests the average was lower than this. The results were for no-moisture machines 16.6% loss, medium moisture 12.8%, and maximum moisture 10.8%. The loss is also affected by the amount of ventilation, as discussed in another place.

Some startling differences in the hatching were secured. In the Utah experiment (1907) maximum moisture produced 329 chicks, medium moisture 319, and no moisture 278 chicks from the same number of eggs put in. At the Oregon Station (Bulletin 100) medium moisture produced 424 chicks, maximum 420, and no moisture 330. In each case the same kind and number of eggs were used in the different machines. Many subsequent experiments showed similar beneficial results from the use of moisture in number of chicks hatched. They showed further that larger and heavier chicks were hatched where moisture was used. These experiments were made with a "moisture" machine.

Amount of Moisture to Use.—The experiments showed that extreme dryness as well as extreme humidity were alike detrimental. The amount to use depends largely upon the amount of ventilation. It was also shown that the range between the temperature of the incubator and the room influenced the humidity of the machine. As the difference increased the humidity decreased. It is this difference or range of temperature between the machine and the room that causes the circulation of air through the machine. As the difference decreases there is less circulation and consequently higher humidity. To maintain a uniform humidity, therefore, account must be taken of the range of temperature, and the supply of moisture governed accordingly. One machine required double the amount of supplied moisture to maintain the same humidity conditions as another machine of different make, due to difference in ventilation. This was shown by the reading of the wet bulb thermometer.

The Wet Bulb as a Guide for Moisture.—The wet bulb thermometer may be used to advantage as an indicator of the proper degree of humidity in the incubator. This is an ordinary thermometer, the bulb of which is covered with a muslin or silk wick, one end of which is inserted in a cup of water. Evaporation is a cooling process, and as the water evaporates on the bulb of the thermometer the temperature is lowered. The rate of cooling depends upon the rate of evaporation. If the evaporation is great or rapid the temperature is lower; if the evaporation is less the temperature rises. The rate of evaporation is influenced by the amount of moisture in the air surrounding the thermometer. The reason for this is that the air takes up moisture fast or slow as it approaches or departs from the point of saturation. When the wet and dry bulbs have the same temperature the point of saturation has been reached and instead of being taken up by the air, moisture is given off; this is what causes rain. The drier the air, therefore, in the machine the greater its thirst for moisture, therefore the more rapid the evaporation of moisture from the wick on the bulb the greater the consequent cooling of the bulb. The drier the machine the more moisture will evaporate on the bulb and the lower the temperature of the wet bulb will be. The more moisture in the air the higher the temperature of the wet bulb will read.

What is the Best Wet Bulb Temperature for Hatching? —In machines with no supplied moisture there was an average wet bulb temperature of 84 to 85 degrees. Machines of the same make, with a tray of wet sand covering half the floor of the egg chamber, averaged about 88 degrees; and in others with a tray of wet sand covering the entire bottom of the machine the temperature was 90 to 91 degrees.

As reported in Oregon Bulletin 100: "An average wet bulb temperature of 87.6 gave 32.6% better hatches than one of 84.5% and slightly better than one of 91%."

Later experiments at the same station confirm these results. The importance of moisture is strongly indicated. The efficiency of the particular incubator used was increased over 30% by using an amount of moisture that maintained a wet bulb temperature of 88 degrees instead of 85. A wet bulb temperature, therefore, of about 88 degrees seems to indicate the best humidity conditions of the incubator.

It is not assumed that it is necessary to supply moisture to the incubator to maintain the proper humidity conditions of the egg chamber. Further experiments showed that different incubators required varying amounts of moisture to maintain the same readings of the wet bulb thermometer, one incubator requiring double the amount of another. The explanation is that the ventilation in the one machine was greater than in the other. The humidity conditions therefore are strongly influenced by the amount of ventilation. Again it was demonstrated that by cutting off the ventilation, the proper wet bulb temperature could be maintained without supplying any moisture.

Moisture and Carbonic Acid Gas.—An explanation, or partial explanation, of the results obtained from supplied moisture was discovered in another experiment. Carbonic acid gas with moisture decomposes calcium carbonate. The egg shell is 93.7% calcium carbonate. Eggs emptied of their contents were put in glass fruit jars, some with water and some without. The jars were then put in an incubator for 21 days, and a strong current of carbonic acid gas was forced through them. At the end of the incubation period, the egg shells in jars containing water in the bottom were broken down or dissolved, while those without the moisture in the jars were unaffected and apparently as strong and hard as at the start. From this it appeared that the shell is something more than a mere covering to preserve the egg. A German experimenter found that there was a loss of lime in the shell during incubation.

What becomes of it? Chemical investigations at the Oregon Station showed that the chick as it was developing within the shell was drawing upon the shell for nutrition;

that there was a gradual increase in the ash or lime contents of the chick, and that the chick was able to break through the shell with strength derived in part from the shell itself. It was also found that the chick contains considerably more lime than the contents of the egg itself. Without the mineral elements of the shell the chick would be unable to grow its frame or skeleton. The shell therefore has a vital function to perform in the hatching process, in the development of the chick. Moisture with carbon dioxide does not merely weaken the shell so the chick will be able to break through, but the dissolved lime of the shell goes to assist in the formation of the chick and determines to a certain extent the future strength and vitality of the fowl.

But is not the moisture within the egg sufficient for this purpose? With certain machines, or those that gave over 30% better hatches with supplied moisture than without, the development of the chick was more complete, there was more ash, phosphorus and protein in the chicks hatched by the machines that were supplied with moisture. The important point brought out here is that the chemical composition of the chick, its strength and vitality, is influenced by the moisture in the machine.

It is highly important that the incubator operator should understand the moisture requirements for best chick development and that he should be able to test the machine for moisture.

There are other methods of determining the amount of moisture in the machine. Thermometer makers have devised different instruments, usually called hygrometers, for determining the humidity of the incubator. Some of these are successful, others are not. When properly constructed they answer the purpose. The percentage humidity as shown by the hygrometer should average about 60%.

These instruments, however, do not give the actual relative humidity in the machine. To get a reading of the hygrometer that will represent the humidity of the whole egg chamber, it is necessary that the bulb be fanned. That is because the moisture which is evaporated from the bulb remains near the bulb, and that part of the egg chamber is therefore more moist than other parts of the machine.

By fanning the bulb, if that were possible in the incubator, the stagnant moisture in the machine would be driven from near the bulb, and the reading would then represent actual conditions of the whole egg chamber. It is in this way that the Government weather bureau observations of the humidity of the air are secured; the hygrometer is fanned. It is not practicable to do this in the incubator, nor is it necessary in practice to know the correct humidity. It is sufficient to know that a certain reading of the wet bulb thermometer, or a certain percentage of humidity as determined without fanning gives the desired condition for successful hatching.

Loss in Weight of Eggs a Guide to Moisture.—Another method of learning whether the humidity is right is to weigh at intervals of a few days during the hatch a number of eggs and note the loss in weight. A dozen eggs may be marked and weighed when put in the machine, then again every six days until the 18th day. An accurate scale of course must be used. Different eggs vary in the amount of loss due to the difference in the texture or structure of the shell, but an average loss within the following limits has been found to be about right:

During the first 6 days..................... 3.5 to 4 % loss
During the second 6 days.................. 4 to 4.5% loss
During the third 6 days................... 4.5 to 5 % loss

Total ....12. to 13.5% loss

(Utah Bulletins 92 and 102. Oregon Bulletin 100.)

After the operator has once tested his machine and learned how much moisture is needed it will not be necessary to continue the use of a moisture test.

Methods of Supplying Moisture.—Moisture may be furnished in the following ways: First by putting a tray of sand under the egg tray and wetting as much of the sand as is necessary. The amount of surface of the sand to moisten will depend upon the make of the machine and upon weather conditions. Second, a pan of water may be put under the egg tray. The sand tray is more satisfactory than the pan inasmuch as the amount can be regulated by increasing or decreasing the area of wet surface. This cannot be done with a pan of water. Third, the eggs may be sprinkled with water at intervals. This method is not very effective. The supply of moisture should be steady throughout the hatch. Fourth, sprinkling the floor and walls of the incubator room is a practice frequently followed. This is not a very effective method of increasing the humidity in the incubator.

Temperature of Incubation.—The proper hatching temperature is about 103 degrees Fahrenheit at the top of the egg. It has been found that this is the average temperature under the sitting hen. During the latter part of the incubation period the temperature is slightly higher than during the first week, due to heat given off by the growing embryo. Good results have been obtained by starting the incubator at $102\frac{1}{2}$ and gradually raising the temperature to 103 at the beginning of the second week; then maintaining it at that temperature. When the chicks begin to hatch the temperature usually rises to 104; that is all right; but it should not be permitted to go higher than 103 for more than an hour or two at a time.

It is very necessary that the temperature be kept steady. A difference of a degree, either higher or lower, for a short time, may not injure the hatch, but the best results can only be secured when the temperature is kept steady. This refers to temperature conditions in all parts of the egg tray. The operator should test the machine by putting a thermometer in each corner and one in the center of the egg tray.

The successful operator is not satisfied merely to hatch a large percentage of the eggs; he hatches them well. Chicks may be hatched but not hatched right. The effect of a wide range of temperature may not show much in the number of chicks hatched, but it will probably show more in the kind hatched, though this fact often escapes notice. The danger from improper temperatures is not so much in the loss of chicks that fail to hatch, as in those hatched with low vitality.

Ventilation.—Not only the incubator room but the incubator itself must have good ventilation in order to carry out the impure air given off by the growing chick within the shell and to supply the required amount of fresh air or oxygen for proper development. The questions of moisture and ventilation are closely related. The greater the ventilation, the more moisture is required. Some incubators are built on the principle that no moisture should be supplied, the proper humidity being maintained by restricting the ventilation. How much ventilation is necessary, is the problem. More data is needed on this point. It has been very clearly established, however, at the Oregon Station, that increasing the ventilation without increasing the moisture reduces the number of chicks hatched and gives chicks of lower vitality. This is due to the extreme dryness of the air surrounding the eggs, not to the greater supply of fresh air. Under such conditions the chick does not make a normal growth. When this excessive ventilation, however, was supplemented with sufficient moisture to maintain the proper wet bulb temperature the results were satisfactory.

Chemical Composition of Chick Affected by Incubation Methods.—It has been definitely shown by the Oregon experiments that moisture and ventilation have a direct relationship to the chemical composition of the chick hatched. The amount of phosphorus, lime, and certain compounds of protein, and even the amount of fat in the chick, is shown to be markedly influenced by the amount of moisture in the machine. Extreme dryness produced chicks that weighed less in dry matter, and lower in protein compounds and phosphoric acid, as well as lime, in tissue. This proves definitely that the conditions or methods of incubation influence not only the number of chicks hatched, but the development and vitality of the chick itself.

Turning the Eggs.—Daily turning of the eggs in the incubator is necessary. This is following nature, for it has been observed that the sitting hen turns the eggs frequently. Unless turned, the yolk rises, being lighter than the white, and the germ spot comes in contact with the shell, or shell membrane, which is fatal. Turning should begin after the eggs have been in the incubator two days.

It is important that the eggs be handled gently. As to method of turning, it is usually well to follow directions of the incubator maker. The eggs should be turned twice a day. Three turnings have given good results.

Cooling the Eggs.—A slight cooling of the eggs each day after the second day is necessary. In cool weather, or when the temperature of the incubator room is lower than 60 degrees, little more cooling will be needed than in the time the eggs are being turned on top of machine. In warmer weather and toward the end of the hatch further cooling is advisable. The amount of cooling given by the sitting hen to the eggs was determined at the Oregon Station by means of a minimum thermometer in a glass bottle filled with water. In 29 tests under five sitting hens the average cooling temperature was 88.9 degrees. The lowest was 81 degrees. This represented the temperature of the interior of the eggs after the hen had been off the nest. Tests made in incubators with the same instrument showed an average of 94.3 degrees, with usual methods of cooling. The tests were made in summer and the eggs were left out of the machine about half an hour.

An instrument of this kind may be used to advantage by the incubator operator as a guide for cooling.

Carbonic Acid Gas and Chick Development.—The hatching of the chick is a most marvelous thing. The shell is the preserver of the egg. Without it the egg could not be kept wholesome more than a day or two. It also protects the germ of the chick from destruction. When it comes to hatching, the shell is an obstacle. The chick before it can hatch and gain its liberty must devise some means by which it can break down this prison wall. It must grow a strong body, and it does it by making use of its enemy, the shell. To break through the shell it must put some of the shell into its body, otherwise it would not be strong enough to hammer down the wall.

How does it turn this enemy into an ally? By another miracle, making use of another enemy—a poison. A poison that would kill the chick—that is deadly to any living animal—the chick uses in obtaining nutriment and in breaking down its prison walls. A death-dealing poison under the operation of this chicken chemistry is made to manufacture something that gives to the chick strength and a right to live. Nature has decreed that fresh air is necessary for the growing embryo, and while the shell protects the contents and prevents the entrance of disease germs and parasites, it admits air. As fresh air enters impure air is given off, and in passing out of the shell these impurities aid in its decomposition; they dissolve certain food in the shell, mineral nutrients, that the chick uses in the growth of its body, and without which it would not come to life. The impurities or poisons are contained in the carbonic acid gas given off by the chick, or exhaled by the lungs. This thing that poisons the body is used by the chick to dissolve the mineral elements of the shell needed in the growth of its body. The shell is honeycombed, so to speak, by this gas that has been cast off by the chick, and the dissolved elements pass into its body. Without the lime extracted from the shell the chick would not have the strength to break through the shell, or breaking through would not have vitality to live.

More than that, the chick utilizes materials in the shell itself for arming itself with a weapon concealed on the point of its beak to puncture a hole in its prison wall and escape.

Carbonic acid gas has an important function to perform in the development of the embryonic life of the chick. A healthy, strong-growing embryo is giving off considerable quantities of this gas, and this gas in passing through the pores of the shell dissolves the necessary minerals for the nutrition of the chick. As the chick grows there is an increase in the amount of carbon dioxide thrown off. Analysis showed more carbon dioxide in the incubator in the later stages of incubation than in the first. Analysis further showed more carbon dioxide under the sitting hen than in the incubator. There was a large amount found under the hen when sitting on glass or china eggs, which led to the discovery that carbon dioxide was being diffused through the skin of the hen's body.

It has not yet been determined whether the same quantity of carbon dioxide must be present in the incubator as under the hen for successful hatching, but experiments at Utah in supplying artificially a large quantity of this gas rather injured than improved the hatching of the incubator. This has been corroborated by experiments at Storrs. It has not been proved that an artificial supply of carbon dioxide will help matters.

Oil on Egg Shells.—Investigation at the Oregon Station (Bulletin 100) has revealed another point in which it is difficult to imitate nature. It was discovered that the shells of eggs under the sitting hens contain a considerable quantity of oil. A fresh unincubated egg contained a small quantity of oil, but eggs that had been under the sitting hen for two weeks contained six times more oil, while the eggs that had been in the incubator contained practically the same amount as fresh eggs, proving that oil was deposited on the eggs by the hen. What the function of this oil is in incubation, is not yet known.

Whatever may be the practical result, it is thus seen that the sitting hen is slowly giving up some of her secrets to scientific research.

CHAPTER XV

ARTIFICIAL BROODING

Artificial Brooding of chicks means furnishing them with proper temperature conditions by either natural or artificial heat during the stage of their growth when the heat of their bodies is not sufficient to maintain life. After they pass the age when they no longer require extra heat, the term brooding does not apply.

The Length of the Brooding Period is about six weeks. The period varies, depending somewhat on the season of the year, or the weather. Successful brooding depends very largely upon so adjusting the brooder temperature that the chicks are able to do without artificial heat at the earliest possible stage. The temperature must be gradually lowered, so that the chick becomes gradually accustomed to colder air.

Brooding Temperature.—The chick is hatched in the incubator at a temperature of 103 degrees. Before it is taken out of the incubator, which should be within a day after the hatch is completed and the chick dry, this temperature should be reduced. The nursery under the egg tray is several degrees colder than the hatching temperature. The incubator should be kept at a temperature of about 100 degrees until the chicks are removed, but extra ventilation should be given. After the chicks are all hatched and dry the door of the incubator should be left partly open to furnish sufficient fresh air. The width of the opening will depend upon the temperature of the incubator room. The best guide is the behavior of the chicks them-

ARTIFICIAL BROODING

selves. The skilled attendant knows from the appearance of the chicks whether or not they have both heat and fresh air enough. If the chicks pant, the temperature is too high or the ventilation is insufficient. On the other hand, if the temperature is too low they will crowd together. It will be found that this is the only safe guide to follow in regulating the temperature and ventilation of the brooders. It is impossible to raise chicks successfully where the chicks are not comfortable at all times, or where the temperature is allowed to go too high or too low at any time.

The brooder should be thoroughly warmed up before putting the chicks into it. The temperature should be about 95 degrees about an inch from the floor before the chicks are put in. There should be a gradual daily lowering of the temperature until at the end of about four weeks it should be about 75 degrees, at which temperature it should be kept for two or three weeks, when artificial heat may be discontinued. There should be always sufficient heat during the day so that the chick when it gets cold may go to the hover or heater and get warmed quickly. If the temperature is too low they will get chilled and crowd together. The safe plan is to furnish sufficient heat so that the chicks will not crowd, but at the same time it is important that they have room to get away from the heat. They should have access to cool air as well as warm air.

The brooder temperature will vary some according to the style of the brooder. With bottom heat the brooder floor will show a higher temperature, but an inch above the temperature should be lower than in a brooder with top heat. But the guide for the attendant should, in all cases, be the comfort of the chicks. If the chicks do not suffer at any time from either too much or too little heat there will be little danger of a high mortality, if other conditions are right, and if chicks of good vitality are put into the brooder. A too high or a too low temperature will result in an inevitable loss of chicks, which may extend over a period of several weeks, and also in weakened vitality in some that live. It should also be remembered that when chicks have once been overheated their powers of resistance have been weakened and they will be unable to stand the same degree of cold as others that have been kept under proper temperature conditions.

Training the Chicks.—In artificial brooding the attendant must do the training that in natural brooding is done by the hen. Success will depend very largely upon the attention given to the chicks during the first two or three days of their brooder life. The chick for the first two days does not know where to find the heat if it should get out from under the hover. It must be taught. By frequently pushing the chicks toward the heat they will soon learn to find it. They should be taught to leave the hover and seek the cooler fresh air occasionally for a time. They should be "kept going" between the heat and the cold for the first two or three days. At any rate, this training should be done several times a day for the first two days, and under no conditions should they be allowed to remain away from the heat until they begin to peep and crowd against each other for warmth. This is fatal. A little time spent, or a little "puttering," for the first two days will save a great deal of trouble and loss later.

Ventilation.—Where fifty or a hundred chicks are kept together in a brooder, that means a great many lungs calling for fresh air or oxygen. They need a little ventilation at the start, but comparatively little to what they require as they grow older. Ventilation as well as temperature must be elastic or progressive. Every day there is an increased demand for fresh air. A mistake is often made in cutting off on the heat as the chicks grow older, instead of increasing the ventilation. The temperature of the brooder may be lowered by increasing the ventilation or admitting more cool fresh air, and it is just as important that the fresh air be increased as that the brooder temperature be decreased.

Floor of the Brooder.—The brooder floor should be covered an inch deep with clean sand. (See "Feeding the Chicks.") In two or three days when the chicks have learned to eat well, a little chaff or cut clover or alfalfa may be covered over the sand for the chicks to scratch in. It is well to begin with the sand.

Sunlight.—Whatever type of brooder is used it should be accessible to the sunlight. Disease germs do not thrive or multiply when exposed to the sunshine.

Types of Brooders.—There are many types of brooders in use, and each year sees a new crop of them. If some of the chicks die, the brooder is very likely to be blamed, and the next year another is tried. There is, as yet, a considerable divergence of opinion as to what constitutes the best type. All the sundry makes of brooders may be divided into the following classes:

1. Individual lamp brooder, indoor and outdoor.
2. Colony brooder.
3. Hot water pipe brooder.
4. Room or stove brooder, with or without hovers.
5. Terra cotta brooder.
6. Electric brooder.
7. Fireless brooder.

The Lamp Brooder.—There are various types of lamp brooders. Some of these are made for outdoors and some for indoors use. There is more or less danger from fire from the lamp, and proper care should be given it to avoid burning the brooder and chicks. It is usually neglect where a fire occurs. Individual lamp brooders range in capacity from about fifty to one hundred chicks.

Colony Brooders.—The colony brooder has many advantages over other systems of brooding. It is not only a brooder but a growing house for the chicks after passing the brooding stage; it is in use most of the year. An individual lamp brooder can be used only for a few weeks, then a house must be provided for the growing chicks. The colony brooding system is less expensive in equipment. A good type of colony brooder is shown in illustration of the Cornell gasolene brooder. This has a capacity of two hundred chicks. The cost for heating by gasolene amounts to from 2 to 10 cents a day, depending on the weather conditions. The cost to build is in the neighborhood of $36.

A colony brooder may be heated by coal oil lamps. Lamps may be placed inside or outside the house and a detachable hover used inside.

A Hot Water Jug may be used to furnish heat in a colony brooder and other brooders. A gallon jug, filled with hot water twice a day will furnish heat enough under a hover 2 feet square for fifty chicks. This is a fireless but not a heatless brooder. Two such jugs and hovers in a colony house 6 x 8 feet will take care of 100 chicks.

A "Fireless and Henless" Brooder.—Sometimes it is desirable to rear hen-hatched chicks in a brooder without the hen. We give an illustration of a home-made brooder on page 325. It is made out of a dry-goods box, a little burlap or flannel, and a gallon vinegar jug. The box may be 3 to 4 feet long, 2½ feet deep and 2½ feet wide, set on edge. Larger sized boxes, however, may be used. The hover should be about 2 feet square, high enough to put the jug under it. Strips of burlap about 4 inches wide are tacked on to the under side of the hover top, which is made of plain matched boards. These strips hang down all over

ARTIFICIAL BROODING

the chicks, not merely around the edges of the hover, and the chicks nestle among the strips of cloth. The jug is filled with hot water and placed underneath in the center of the hover. If the water is too hot a little felt should be wrapped around the jug to avoid burning the chicks. This will also help to retain the heat longer.

During the first week, the jug should be refilled night and morning. As the chicks grow older, once a day will be

HOT-WATER JUG BROODER
A dry goods box, a hover and a gallon jug of hot water.

often enough. In warm weather the jug may be dispensed with when the chicks are two weeks old. Later, the hover may be removed and the box used for brooding the chicks till they are about three months old. This brooder will take care of fifty chicks easily. They should be given more room when two or three months old. In such a brooder, with felt or burlap strips hanging over the chicks, it requires very little artificial heat to keep them warm, as they keep themselves pretty comfortable when all are close together under the hover. Such a brooder may be used for both hen-hatched and incubator chicks. Fireless brooders of this kind will rear chicks successfully, but a fireless brooder without heat of any kind is not practicable. If the weather is not too cold and the chicks have close attention the heatless brooder may raise chicks successfully, but the labor cost is great.

Long Hot-Water Pipe Brooders.—Where large numbers of chicks are hatched at one time this is probably the most convenient way of brooding them. A heater is located at one end of the house, and from this hot-water pipes run

CONTINUOUS BROODING SYSTEM
(Courtesy Kansas Experiment Station.)

the length of the house under hover but over the chicks. A brooder house a hundred feet in length or more may be heated in this way. This method has been largely used in the production of broilers. The main objection to this system is the cost of equipment. Where the purpose is to raise pullets for fall laying, the chickens have to be transferred to other houses later.

Experience has shown that it is not practicable to use this type of brooder house as a growing house for pullets. Pullets do better when put in open-air houses and given free range. With a brooder house of this kind it will pay to transfer the chickens to free range in open houses, where the purpose is to produce breeding stock or layers. When this is done the cost of producing the mature pullet is greater than is necessary. The colony brooding house system is more economical.

Stove or Room Brooding.—This system is used where large numbers of chicks are hatched. A specially con-

ROOM OF STOVE BROODING—A NIGHT SCENE
At a certain radius from the stove the chicks find the proper temperature. (Oregon Station.)

structed oil-burner stove set in the center of a room about 20 x 20 feet, furnishes the heat for 1,000 chicks or more. Sometimes a coal-burning stove is used. There are no partitions or hovers. The heat is regulated so that at a distance of about 2 feet from the stove, and in a circle around it, the chicks find a proper brooding temperature. Closer it is too hot, farther away, too cold. At night the chicks cover a space of about 12 inches wide and 2 feet from the stove. During the day they run all around the room and get exercise in that way.

During the first few nights a "fence" 12 inches high, made of poultry netting covered with burlap, is placed in a circle around the stove, but several feet farther away from the stove than the chicks. This keeps the floor a little warmer and prevents the chicks getting into the corners of the room. Among the advantages of this system is convenience of operation and saving of labor. The oil is fed

A STOVE BROODER WITH HOVER

automatically to the burner from a tank on the outside wall, and it requires very little attention. Occasionally the soot clogs up the burner, and sometimes the stovepipe will burn out, with some danger of setting fire to the house. Should the heat for any reason be shut off, there will be a large mortality where so many chicks are kept together. This system has not given the results in growth of chicks that is desirable in a good brooding system.

A Room Brooder with Hover.—A modification of the room brooding system is to substitute for the stove in the center of the room a gasolene or distillate heater placed in a lean-to and lower than the floor. From this a hot-air flue conveys the heat under the floor and up through an opening in the center of the room. A hover 6 or 7 feet square is set in the center of the room and over the hot-air inlet. This system is more satisfactory than the stove heater without the hover and is much more economical in

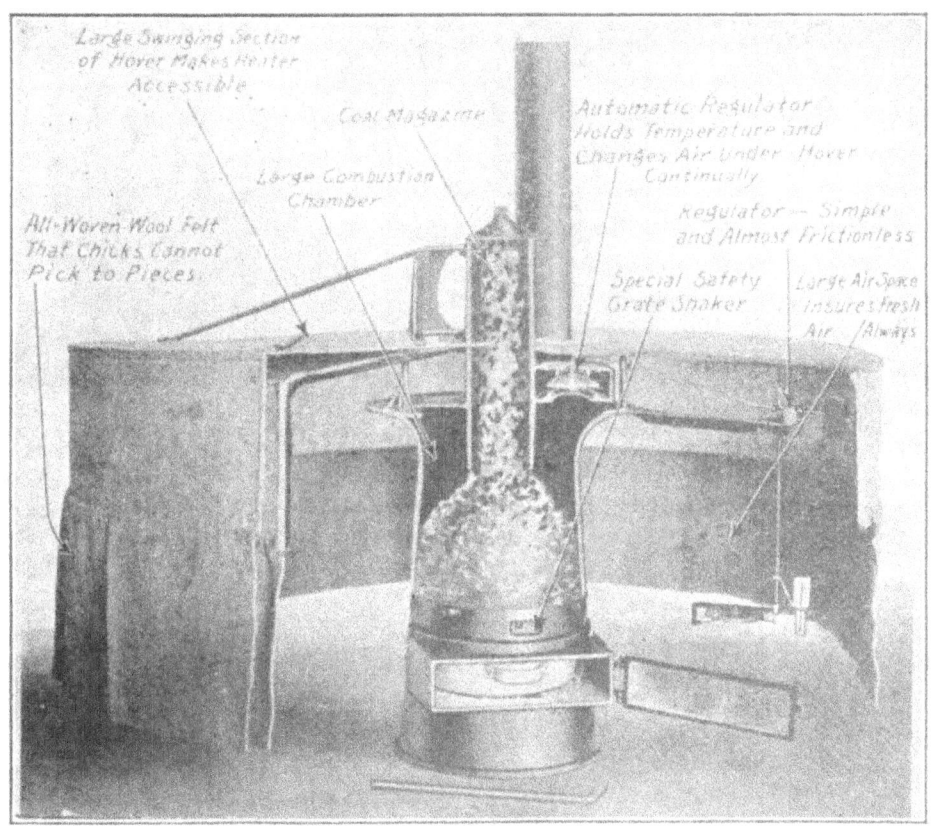

A STOVE BROODER, SHOWING HOVER AND DIFFERENT PARTS OF BROODER

fuel or oil. A thousand chicks may be brooded in such a brooder house with this system of heating and hovering.

The main contention, however, remains unsolved as to whether it is possible to raise chicks as successfully in flocks as large as 1,000 in one room as in smaller flocks. It is undoubtedly true that as the numbers in the flock increase the tendency is for the mortality to increase and for

ROOM BROODING, WITH OIL OR GAS HEATER, OUTSIDE OF ROOM IN A LEAN-TO

Heat comes up through the floor, under the hover, which is shown raised in the picture.

FLOCK OF 8,000 YOUNG PULLETS
Farm of J. W. George, Petaluma.

ARTIFICIAL BROODING

a retarded growth. Whether the difference is great enough to warrant the extra labor cost involved in keeping them in small flocks, is open to doubt. Where several thousand chicks are hatched, the old system of keeping them in flocks of fifty is hardly practicable on account of the labor cost as well as equipment cost.

Brooding Systems in General.—It should be understood by the poultryman that the brooder is not always responsible for chick mortality. The chicks die sometimes in spite of the brooder. If they do, the poultryman should not conclude at once that the brooder is wrong and proceed to purchase or build a new one. It has been abundantly demonstrated that the

CORNELL GASOLINE BROODER

chicks often come to the brooder with vitality so low that it is impossible to raise them. The trouble may have been in faulty methods of incubation, or it may be traced back to a lack of vigor in the breeding stock that laid the eggs. A simple experiment of setting a few hens at the time the incubators are set, using the same kind of eggs and brooding the chicks naturally, will show whether the fault was in the breeding stock or not.

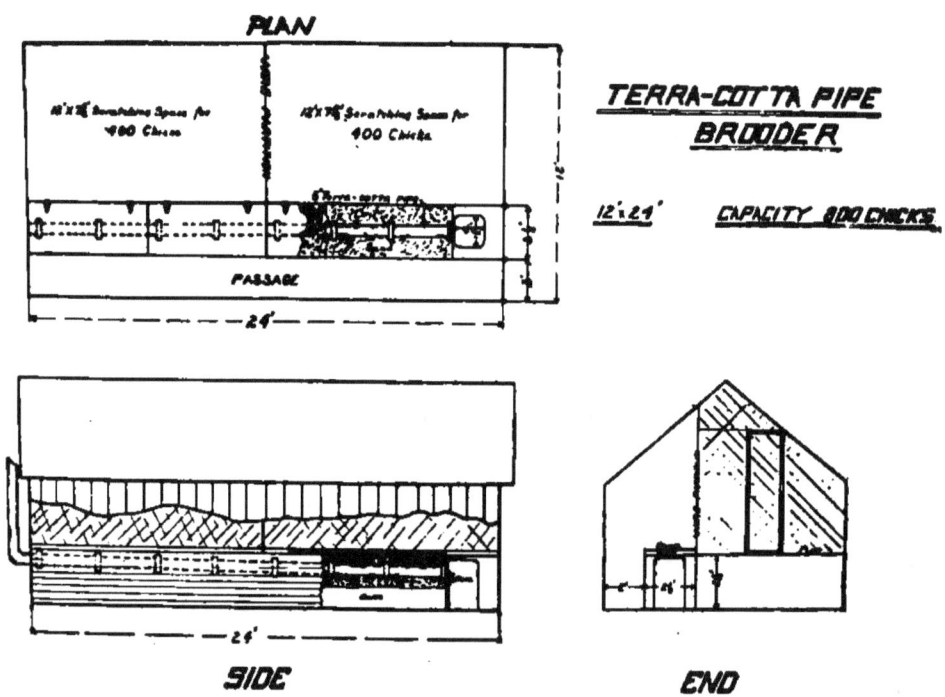

TERRA-COTTA BROODER

This brooder has been largely used by the Petaluma poultry raisers. Besides being cheaply constructed, it is economical in fuel.

CHAPTER XVI

MARKETING EGGS AND POULTRY

During the past few years considerable attention has been directed to the importance of improved methods of marketing poultry products. This is a subject that concerns the consumer as well as the producer. To the former it is a question of how to get eggs and chickens of good quality and at prices he can afford to pay. To the producer it is a question of handling the eggs and marketing them in such a way that he can get a satisfactory profit. Improved methods of marketing would mean a larger business, as well as a more profitable one, because there would be a much greater consumption of eggs and chickens if the consumer could always be assured of their good quality. The poultry-keeper is vitally interested in anything that will increase the consumption of poultry and eggs and also in anything that will enable him, through better handling of the product, to get it to the consumer in a condition that will bring him a higher profit.

The poultryman must not give less attention to production, but he must give more earnest attention to efficiency in marketing if he would not lose what he may have gained in efficiency in production. It is one thing to produce the chickens and the eggs, it is another to get value for them. There is a pleasure in producing a superior article, but the producer measures his success finally by the market returns.

MARKETING EGGS

Two factors that the poultryman must consider in seeking a good market for his eggs are quality and quantity, in other words the two Q's. If he has both Q's there is no reason why he may not develop a special market and reap a special profit. A few cents difference in the price received for his eggs may not amount to much in the course of a week or a month, but in the course of a year, if he has only a hundred hens, the difference between a good and a poor market would amount to probably $50, or 50 cents per hen; or $500 on a thousand hens. This represents an increase of 20 or 25%. The great majority, however, of the producers do not keep one hundred hens. The average farm flock is about 50 hens, and there is not much incentive for the owner to work out better marketing methods.

In the first place he has not the quantity of eggs to make regular shipments. He cannot go to a dealer or retailer or to private consumers and guarantee a certain number of cases a week or twice a week. He may produce the quality, but without the quantity it will not profit him much. On the other hand he may have the quantity, but not the quality, and he will lose a large part of his profits. The producer cannot expect the highest prices unless he has in addition to quantity the quality that will command them.

The problem of marketing is a simple one to the producer who has a large flock, but to the small producer it is a difficult one, though the small producer in the aggregate furnishes the great bulk of the poultry products of the country. The producers are suffering a loss in marketing the product of probably 20 to 25%, or anywhere from fifty to a hundred million dollars a year in the United States, because they individually have neither the quantity nor the quality to interest them in better marketing methods. They lack the two Q's. Is this loss inevitable? Not necessarily.

Where is the remedy, if there be any? In the first place, the remedy is largely educational. The trouble has been located and the remedy is known; it is a question of applying it. A study must be made of egg quality. The difference in quality of eggs and the factors affecting the quality must be clearly demonstrated to the producer and consumer alike. When the former understands fully that the egg is a perishable product and that 15 to 20% of the real value of the egg is lost in the handling under present marketing methods; in other words, that the quality of the egg varies to this extent, it will not be very long before the producer sets about rectifying this great marketing blunder. He will set about improving the quality.

In the next place there must be developed a greater community of interest among the producers, in other words, co-operation. If the business is going to continue to be a business of small producers it is imperative that they get together to the extent at least that regular guaranteed shipments may be made in sufficient quantity, if the direct method of selling is to be followed.

If, on the other hand, the indirect method is to be followed, that is, selling the product through dealers instead of shipping direct to consumers, then the producers should co-operate to the extent of compelling the dealers to modify or reform their methods of buying. Egg production is a part of a well regulated system of diversified farming, and if this system of farming is to be permanently successful there must be a greater community of interest developed.

Indirect Selling.—Two distinct methods are followed in marketing poultry products. They may be called direct and indirect. The indirect route varies somewhat in length or efficiency. Shipping to the retailer comes nearest to the direct method in efficiency. The most indirect way is where the producer sells to the huckster, the huckster to the local store or shipper; the latter to the commission man in the city, who sells to the jobber, and the jobber to the retailer, the eggs finally reaching the consumer from the retailer. That means five middlemen. Some of these middlemen, however, have been eliminated in many markets.

By far the largest proportion of eggs reach the consumer by the indirect route. The reason is largely because the community of interest or co-operative spirit is not highly developed among the farmers or producers. It is fair to say that during the past two or three years considerable progress has been made in improving the service and lessening the cost of marketing eggs by the direct as well as indirect method. The business of buying eggs on commission and the abuses that followed have been largely eliminated in most of the large markets. The bulk of the business is now done by large jobbers. The service has been improved and the cost of marketing reduced. Many of these jobbers have large storage plants, which afford an outlet for surplus stock during the season of plenty.

How the Costs are Added.—The tabular exhibit on page 337 shows how, through the indirect method, the costs are added to the egg in New York City before it reaches the consumer. It is taken from a report of a State Food Investigating Commission.

The cost of marketing is here shown to be over 50% of the amount received by the producer. This would not apply to all markets; the cost would be less in some. In a western state on July 30, farmers were receiving 22½ cents a dozen in trade from the local stores, or 21 cents in cash. The local stores received 22½ cents from the jobbers after paying express charges to the city one hundred miles distant. The retailer paid the jobber 27 cents, and the retailer sold the eggs to the consumer at 30 cents to 35 cents, depending on the grade. Good eggs that the producer received 21 cents for were sold to the consumer for 35 cents. Eggs from another producer that were not as good were paid for at the same rate, 21 cents, but were finally sold to the consumer for 30 cents.

ANALYSIS OF RETAIL PRICE OF EGGS IN NEW YORK CITY

Per dozen

Producer's price	$0.20	$0.20
Shipper's charges:		
(a) Labor in collection and packing	.005	
(b) Cases, fillers, and packing	.0073	
(c) Transportation charges to city	.0106	.023
Commission for handling	.01	.01
Jobber's charges:		
(a) Cartage from dock to store	.00133	
(b) Candling and grading	.00666	
(c) Storage and insurance	.016	
(d) Jobber's profit and charges	.01	
(e) Delivery to the retailer	.004	.038
Retailer's charges:		
(a) Operating expenses, 10%	.0271	
(b) Retailer profit, 5%	.01497	.042
Price paid by consumer		$0.313

By direct marketing is meant selling direct to the consumer by the producer, or an association of producers, and indirect marketing is the method followed where the eggs pass through the hands of one or more middlemen after leaving the farm and before reaching the consumer.

Direct Selling.—The most profit will be made by shipping direct to consumers, in this way eliminating all middlemen charges. The producer should endeavor to establish a trade with city customers. It is possible to furnish the consumer eggs of first quality at prices lower than he pays for eggs of the same quality through the regular channels of trade, and at the same time secure a better price than he could otherwise secure.

This was demonstrated by the poultry department of the Oregon Agricultural College. For better eggs than he formerly got for 35 cents a dozen the consumer paid the producer 27 cents, plus express, which was about 3 cents a dozen. This was a gain of 6 cents to the farmer over the usual method of selling to local stores, and a saving to the consumer of 5 cents a dozen. The eggs were shipped in crates of 12 dozen. Some of the customers divided the eggs with their neighbors; others used them all, though they had to keep some of them two or three weeks. At the end of that time they reported that the eggs were better than those purchased at stores. This is a trade worth looking after.

A 12-DOZEN CRATE

This crate may be used for shipping eggs to consumers by express or parcel post, direct from the farm.

The disadvantages of direct selling are, first, that the express charges operate against shipping in small quantities. The express charges on a 12-dozen case may be about as much as on a 30-dozen case, and two cases of 30 dozen may be shipped at the same cost practically as one case. If the express charges were to be fixed at so much a dozen, irrespective of the size of crate or number of dozens shipped, it would encourage the extension of direct shipment. As now, the rates make it an object for the farmer who has good eggs to sell to take them to the local shipper at the same price another farmer gets who has poor eggs to sell, because the local shipper can ship the eggs at less cost than the individual farmer on account of being able to make larger shipments.

Second, there will be times when the producer will have

MARKETING EGGS AND POULTRY

a surplus of eggs that he has no regular customers for; then there will also be culls. For these he must find another market. These would usually have to be marketed through the indirect way. Third, there there would be the cost of collection, or the probable loss from bad accounts. By requiring a bank reference, or other satisfactory reference, there would be little probability of loss.

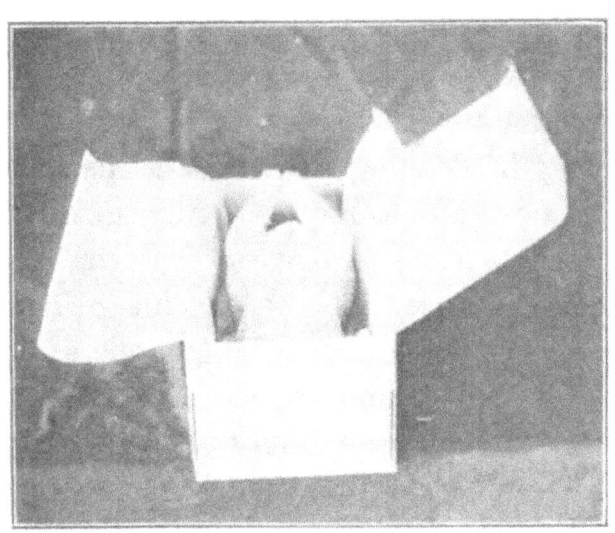

A ROASTER
Showing method of packing in a parcel post package.

Selling to City Retail Stores.—Retail stores offer a market that should not be overlooked. Shipping direct to retailers comes near being direct selling. The producer can ship in larger quantities than he can to private customers and obtain a better express rate. Retail stores that have a fancy trade will pay a premium for fancy eggs. They have the marketing machinery all ready running. They attend to deliveries and collections. The poultryman who can guarantee regular case shipments of high quality stock will

A PARCEL POST PACKAGE
Showing eggs wrapped.

often do as well by shipping to the retail stores as in any other way.

Hotel and Restaurant Trade.—There is no reason why the producers should not be able to furnish the hotel and restaurant trade where they have the quantity and quality of eggs. Why it is not now done more largely is due primarily to a lack of business management on the part of producers more than anything else. If the producer has the quality and quantity he should go to the best hotels and restaurants and endeavor to find a market.

Parcel Post Shipments.—It has been fairly well demonstrated that market eggs may be shipped successfully by parcel post. By a recent ruling of the post office department packages weighing as much as 50 pounds may be sent by mail. The main objection to parcel post shipments has been that the cost of the container or parcel has been too great to admit of a profit being made. Different manufacturers, however, have been at work on the problem, and containers may now be obtained at prices that are within reach. It is not expected that a profit can be made by the producer unless he can get a little higher price for the eggs than is paid for second grade eggs in the city. The producer in working up a parcel post trade must cater to that class of consumers who wish eggs of superior quality and are willing to pay more for them than for eggs of inferior quality. The consumer is able to secure in this way eggs

THE CARRIER

The rural mail carrier takes the eggs from the farm to the post-office.

of first quality at the same price he pays for eggs of poorer quality. There will, however, always be another class of consumers who are unable or unwilling to pay the price that will enable the producer to ship them by parcel post.

The post office department has made certain regulations in regard to shipments that the shipper must observe. The package must be made in such a way that the contents of the egg, if broken, will not run out of the package and injure other mail matter. To obviate this it is required that each egg be wrapped separately, except when in packages exceeding 20 pounds; those are not required to be wrapped. All parcels must be labeled EGGS. Parcels weighing more than 20 pounds will be accepted, but the crates or boxes must have tight bottoms. Such packages must be marked "Eggs—this side up." They will be transported outside mail bags. Producers wishing addresses of manufacturers of shipping packages, may apply to their home state experiment stations.

The larger the package or the more eggs shipped in one package, the lower the cost per dozen for parcel post. For instance, in the first and second zones the first pound costs 5 cents, while each additional pound up to fifty, costs but 1 cent. A twenty-pound parcel would cost 24 cents or 1.4 cents a pound. Five dozen eggs weighing, with container, ten pounds, will cost for postage in a distance from 50 to 150 miles, 14 cents, or 2.8 cents a dozen. The return postage would be 6 or 7 cents. The postage should be charged to the customer, likewise the cost of the container. When the container is returned the customer would be given credit for it.

To make the business a success there must be mutual coöperation between the producer and consumer. It must be understood by the producer that the consumer will purchase his eggs only so long as he can furnish a superior article.

On the other hand, it will be understood by the consumer that he will secure from the producer better eggs than he can secure through indirect channels at the same price. It means a better profit to the producer and at the same time a saving to the consumer.

The producer should be fully alive to the possibilities of this method of selling and should take particular care to grade his eggs as to size and color, separating the white and the brown eggs and discarding all under-sized and over-sized, ill-shaped and dirty or stained eggs. Above all, he must be sure that he ships nothing that is not perfectly fresh. He will not long retain his customers unless he gives heed to those points. Again, it will pay well to use neat and clean packages and also a wrapping-paper of proper size and quality. By treating the customers fairly and pleasing them he will be able to secure others because a pleased customer will recommend others to him.

In regard to fixing the price, probably the most practicable method is to have an agreement with the consumer that the price will be the highest wholesale quotations in the daily papers plus so many cents premium. This should be sufficient to pay the postage and package, and the cost of the extra care given the eggs.

Comparison of Direct and Indirect Selling.—As between the two methods of selling—direct and indirect—the former undoubtedly favors the maintenance of the higher standard of quality. In shipping to consumers the producer is directly responsible for his product. Any complaint will come to him direct. He is able to retain his customers only so long as he furnishes eggs of superior quality. He has a direct interest in the quality of his product. By the indirect method the producer's identity is not known. When his eggs are marketed with those of a hundred other farmers, there is no particular object in taking pains to preserve the quality. Second, by the direct method the producer of good eggs is able to get a price for them which he is not able to secure when his eggs are marketed in common with those of various other producers. Third, by shipping direct to consumers he is able to add to his profits a part at least of the profit that went to various middlemen. There will always be needed, however, an outlet for surplus stock which is only furnished at present through the medium of the jobber and retailer.

Buying by Quality.—This brings up the question of buying by quality. The greatest objection to present methods of marketing eggs is the heretofore almost universal practice of dealers paying for them by the dozen without reference to their quality. There is one price for eggs at the local stores. A farmer who once a week gathers his eggs from stolen nests under the barn and in the fence corners and takes them to town, gets the "going price" at the store. Another farmer who gathers the eggs regularly from clean nests once a day and twice a day in hot weather and takes them to town every two or three days, gets the same "going price." He has no inducement to maintain the good quality of his eggs. The system does not encourage it; rather it encourages carelessness on the part of the producer. It offers a premium on dishonesty. The wonder is, on the one hand, that the producer is able to maintain his integrity, and on the other that the consumer is able to get an egg of good quality.

Before the eggs reach the consumer the broody hen sits on them a while, the sun incubates them a while, the railroad rides on them a while, the city storekeeper broods over them a while, and the consumer raves over them quite a while.

The storekeeper is not alone responsible for this method. The dealers and commission men follow the same method in buying from the storekeeper. They buy them by the case-count. It is called the case-count system. During the past few years an active campaign has begun in several states to do away with the case-count system and substitute a system of buying on the basis of quality. This means that the dealer in purchasing a case of eggs candles them, computes the loss due to shrinkage, blood-rings, etc., and pays accordingly. This is known as the "loss-off" method of buying, which is really paying according to quality. If this system comes fully into vogue it will result in saving millions of dollars a year to the producers, for the loss finally is charged up against the producer. It will also fasten the attention of the producer on the importance of breeding for size of egg, feeding for quality in the egg, and on methods of handling the egg that will best preserve its quality.

Grades of Eggs.—Under the old system, an "egg was an egg," and at the present day, in the majority of primary markets, one egg is as good as another. Now before eggs reach the consumers in large cities they have to stand an examination, and the expert finds that there are various kinds of eggs, and a name or grade is given to each kind. The different grades of eggs recognized by expert candlers in the large markets are described in Bulletin 160, Bureau of Animal Industry, as follows:

Fresh Egg.—An egg to be accepted as a first, or fresh egg, must be newly laid, clean, of normal size, showing a very small air cell, and must have a strong, smooth shell of even color and free from cracks.

Checks.—This term applies to eggs which are cracked but not leaking.

Leakers.—As indicated by the name, this term applies to eggs which have lost a part of their contents.

Seconds.—The term "seconds" applies to eggs which

have deteriorated to a sufficient extent as to be rejected as firsts. The several classes of eggs which go to make up this grade may be defined as follows:

(a) **Heated Egg.**—One in which the embryo has proceeded to a point corresponding to about 18 to 24 hours of normal incubation. In the infertile egg this condition can be recognized by the increased color of the yolk; when held before the candle it will appear heavy and slightly darker than in the fertile egg.

(b) **Shrunken Egg.**—This class of seconds can be easily distinguished by the size of the air cell. It may occupy from one-fifth to one-third of the space inside the shell.

(c) **Small Egg.**—Any egg that will detract from the appearance of normal eggs on account of its small size will come under this class, although it may be a new-laid egg.

(d) **Dirty Egg.**—Fresh eggs which have been soiled with earth, droppings, or egg contents, or badly stained by coming in contact with wet straw, hay, etc., are classed as seconds.

(e) **Watery Egg.**—Those in which the inner membrane of the air cell is ruptured, allowing the air to escape into the contents of the egg, and thereby giving a watery or frothy appearance.

(f) **Presence of Foreign Matter in Eggs.**—Often eggs are laid which show small clots of blood about the size of a pea. These are sometimes termed "liver" or "meat" spots.

(g) **Badly Misshaped Eggs.**—Eggs which are extremely long or very flat, or in which part of the shell's surface is raised in the form of a ring; in other instances a number of hard, wart-like growths appear on the outside of the shell.

Spots.—Eggs in which bacteria or mold growth has developed locally and caused the formation of a lumpy adhesion on the inside of the shell.

Blood Rings.—Eggs in which the embryo has developed to a sufficient extent so that it is quickly recognized when held before the candle.

Rots.—Eggs which are absolutely unfit for food. The different classes of rots may be defined as follows:

(a) **Black Rot.**—This is the easiest class of rots to recognize and consequently the best known. When the egg is held before the candle, the contents have a blackish appearance, and in most cases the air cell is very prominent. The formation of hydrogen-sulphid gas in the egg causes the contents to blacken and gives rise to the characteristic rotten-egg smell, and sometimes causes the egg to explode.

(b) **White Rot.**—These eggs have a characteristic sour smell. The contents become watery, the yolk and white mixed, and the whole egg offensive to both the sight and the smell. It is also known as the "mixed rot."

(c) **Spot Rot.**—In this case the foreign growth has not contaminated the entire egg, but has remained near the point of entrance. Such eggs are readily picked out with the candle, and when broken show lumpy particles adhering to the inside of the shell. These lumps are of various colors and appearances.

White and Brown Eggs.—The color of the shell in certain markets affects the price of the egg. Most markets in the United States prefer the white egg. New York markets pay a premium for white eggs. Among the best grades, brown eggs sell for about 20% less than white eggs. There is no difference, however, in price of cheaper grades on account of color. In San Francisco the brown egg is also discounted. In Boston the reverse is true, the brown egg being preferred.

This does not mean that there is any difference in the quality of the brown and the white egg. The difference in price is, however, undoubtedly due in part to the mistaken notion that there is such a difference in quality. The color of the shell has nothing to do with the quality of the contents.

Classification of Eggs.—The classification of eggs is controlled by city mercantile bodies interested in the buying and selling of farm produce. The classification varies in different cities. The average producer knows no classification; in other words, on the farm "eggs is eggs"; but by the time they reach the city markets there is a rigid culling and they are separated into many grades or classes. A study of these grades and classifications indicates that the losses occur largely through wrong methods of handling the eggs before they leave the farm, and the producer must be the loser in the end.

The following, taken from the *New York Times* of May 1, 1914, shows the many different classes into which eggs are divided by dealers in that city and the range of values placed upon them:

Fresh gathered extras, 23, 23½ cents.
Storage packed firsts, 22, 22½ cents.
Regular packed firsts, 21½, 22 cents.
Seconds, 20½, 21 cents.
Thirds and fourths, 19½, 20 cents.
Number 1 dirties 19½ cents.
Number 2 dirties 16, 18 cents.
Checks good to prime, dry 18, 19 cents.
State, Penn. and nearby hennery, whites fine to fancy 24 cents.
Gathered whites, fine to finest, 23½ cents.
State, Penn. and nearby whites, fair to good, 22, 23 cents.
Western gathered whites, 22, 22½ cents.
State, Penn. and nearby hennery browns, 23, 23½ cents.
Gathered brown and mix colors, 21, 22 cents.
Baltimore selected, 22, 23 cents.
Western, 20, 22 cents.
Tenn. and other good Southern, 19, 21 cents.
Far Southern, 17, 18 cents.

The following quotations from the *Chicago Herald* of August 23, 1914, gives the grades recognized in that city and the relative values placed upon different grades. The range on that date was from 27 to 15 cents a dozen:

    Extras .... 26, 27 cents
    Firsts .... 22, 23 cents
    Ordinary firsts .... 19½, 20½ cents
    Checks .... 15, 16½ cents
    Dirties .... 16, 17½ cents
    Miscellaneous lots............... 16½ to 22½ cents

Conditions that Injure the Quality of Eggs.—While the new-laid egg is "one of the most delicious morsels to the human palate and one to fill the heart of man with loving-kindness," it should be clearly understood that under certain conditions it rapidly loses its peculiar excellence. It is a perishable article. The rate of deterioration is influenced by many things, such as:

Insufficient Nests.—A new-laid egg may not be fresh; that is, it may have lost its freshness by the time it is gathered in the evening if there are insufficient nests for the hens. If hens are continually on the nests throughout the day the embryo may begin to develop in the fertile eggs. At any rate, they will have lost some of their freshness. When the nests are crowded all day it is a sign that there are too few nests.

The Broody Hen.—At certain seasons of the year the broody hens are responsible for a considerable loss in the quality of the eggs. If permitted to remain with the flock of layers the broody hens will injure a great many eggs by sitting on them and starting incubation.

Stolen Nests.—Rotten and stale eggs often come from the stolen nests. They are found by the children under the corn-crib, in the straw or hay-stack or fence corner. When these "finds" are mixed with the regular supply of fre eggs and the city consumer gets one on his breakfast table, it will be "no more eggs for him," and the consumption of eggs is curtailed. With proper nesting arrangements the hens will not be so likely to steal their nests.

Dirty Nests.—Dirty nests affect the flavor and keeping qualities of the egg. Germs of decomposition may enter the egg; dirt stains on the egg will spoil it for select trade. The nests should frequently be examined and the nesting material renewed. Clean, fine hay or straw in the nest will help to keep up the quality and grade of the eggs.

Fertility of Eggs.—The starting of incubation or the development of the embryo starts most of the trouble between producer and consumer. If there were no males in the flock and the eggs were not fertile there would be fewer complaints of bad eggs. It has been estimated that the loss in quality of eggs due to the presence of males in the flock amounts to millions of dollars a year in this country. Males are necessary to fertilize the eggs for hatching, but not for any other purpose. The infertile eggs have better keeping qualities in warm weather than the fertile. There is no difference in the egg yield whether the males run with the flock or not. They should be removed from the yards after the breeding season is past. If not desired for breeding in another season the males should be marketed. Keeping them till fall only means that a lower price will be received for them. If, however, eggs are kept in a cool place, or at a temperature low enough to prevent germ development, it will make no difference in their keeping qualities whether fertile or not.

Gathering the Eggs.—In cold weather if the eggs are left in the nest over night they are liable to freeze, and in warm weather if they are not regularly and frequently gathered there is likely to be germ development. In certain sections of the country where the temperature frequently reaches a point where incubation will begin, the eggs must be frequently gathered. The rule for gathering the eggs should be, once a day, except in warm weather when it should be twice a day.

Storing or Keeping the Eggs.—Cooling the eggs checks deterioration. After being gathered they should be kept in a moderately cool place until shipped. The best temperature is between 45 and 60 degrees. A cool, dry cellar is the best place. Objectionable odors may pass through the shell to the contents. The eggs should not, therefore, be stored near decaying vegetables, coal oil, or other things that may injure their flavor.

Shipping the Eggs.—The eggs should be shipped to market as frequently as possible, at least once a week, and in warm weather twice a week. The fresh egg soon becomes a stale egg.

Clean Eggs.—To grade as first quality the eggs must be clean. The hen covers the egg contents with a clean shell to preserve its purity; the poultryman should be as careful to keep its exterior clean. Dirty nests and dirty yards cause the dirty eggs. Dirty yards mean dirty feet and dirty feet mean dirty eggs, and dirty eggs mean loss of profits.

Washing the Eggs.—Washing the eggs may injure their keeping qualities and spoil their natural appearance. It is better, however, to wash the eggs than to market them dirty. Rubbing with a clean, moist cloth may be all that is necessary. A little washing powder or sapolio may be used where necessary.

Taking the Eggs to Market.—Sometimes the quality of the eggs is injured on the way from farm to town. If the sun on a hot day in a long drive strikes on the eggs it will injure them seriously. The sun is an incubator, and it is not well to incubate eggs that are intended for the breakfast table of some city customer who believes in pure food and good living. Undue jarring of the eggs should also be obviated.

Exposing for Sale.—A great many retailers treat eggs as though their quality was improved by warmth. In cold weather the eggs have a place near the stove very frequently, and in summer they are put in the window where they can get the benefit of the sunshine. This treatment should be reversed. Displaying eggs in the window where the hot sun strikes on them is not a good advertisement for the eggs.

Grading Eggs.—In large markets eggs are graded according to size, color, and quality. Consumers must have some assurance of the quality of the eggs, otherwise they will be afraid to eat them. The grading of eggs, therefore, by assuring purchasers of their quality, tends to increase consumption and the profits of the producer.

1. Size.—The poultryman should breed for size of egg as well as number. The importance of this has not been brought home to him very strongly, because his eggs have, in most cases, been paid for by the dozen and not by weight or size. In the future the size of eggs must be reckoned with. Whether they will be sold by the pound or the grade, the larger eggs will command the higher prices. This is now the rule in many of the leading markets. Poultrymen should not be satisfied until their flocks produce eggs that average two ounces per egg, or $1\frac{1}{2}$ pounds per dozen. Hens laying smaller eggs should not be used as breeders. Uniformity in size has also a market value. The very large egg as well as the very small egg should not be used for the special trade. The more uniform the size the better the eggs look. Care and feeding of the fowls have an influence on size of egg, as explained in Chapter XI.

2. Color.—Uniformity in color has also a market value

as shown by market quotations in several of the large centers. The preference of consumers for eggs of a certain color is based on a fad, but so long as the fad does not interfere with his business the poultryman will take cognizance of it and endeavor to furnish eggs of the color that will command the highest price, whether they be white or brown. In many markets no importance is placed on color, but if the poultryman has eggs of different colors it will pay him to separate them, filling one end of the crate with white and the other with browns if he has not enough to make a case of each, or if cartons are used they may be separated in this way.

3. **Shape of Egg.**—Abnormal eggs should be culled out. These include double-yolked eggs, ill-shapen eggs and soft-shelled eggs. With proper attention to breeding and management of the fowls the percentage of culls should be small.

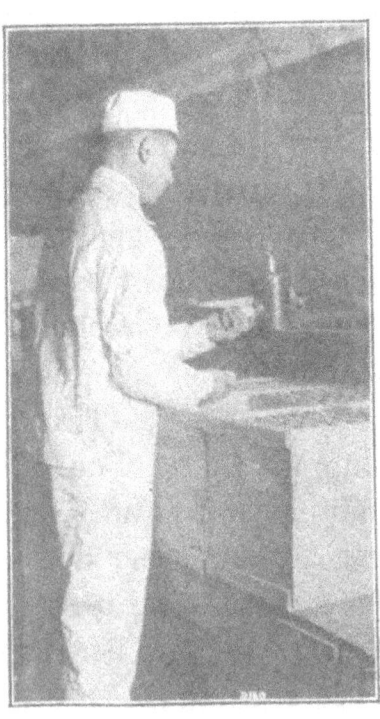

COMMERCIAL EGG CANDLING

(Courtesy, Bureau of Chemistry, U. S. Department of Agriculture.)

4. **The Egg Contents.**—Eggs that are either too highly colored or too pale in the yolk are objectionable. The color is controlled by the feed. (See Chapter XI.)

5. **Candling.**—The quality of the egg contents is determined by candling. By candling it is meant that the egg is subjected to a light that reveals, for all practical purposes, its real market value. There are various methods of candling eggs. In early methods a tallow candle was used, hence the origin of the name candling. A candle was put inside of a

small box in which a hole, a trifle smaller than an egg had been cut, and by holding an egg at the opening the condition of the egg could be seen. The candling is done in a reasonably dark room. Instead of a candle an ordinary oil lamp may be used, placing the lamp in a box. A tin chimney, with a hole in the side, may be

A GOOD EGG TESTER

A kerosene lamp set inside of a shoe box or a cereal box, with a hole opposite lamp flame, makes a good tester.

put on the lamp in place of the glass chimney. Special testers or chimneys of this kind may be purchased at poultry supply houses. An electric light bulb may be used inside of an ordinary shoe box, or other box. Expert candlers usually use a tester having two holes so that he can take two eggs up at one time, one in each hand. A length of stovepipe, with two holes in the side and an electric light bulb inside, is frequently used. An expert candler can test 1,000 dozen eggs in a day.

Consumers Should Candle Their Eggs.—If the consumer wishes to assure himself, before eating the eggs, that they are all right, he may very readily do so by using a small shoe box or breakfast cereal box with the electric light. It should not consume more than a minute to candle a dozen or two dozen eggs in this way. He may detect the bad egg or determine the quality of the egg without first breaking it on to his breakfast toast. He can also check up on the grocer or farmer as to the age of the egg, and assure himself that he has not purchased a stale egg under the name of new-laid. If the eggs are not up to the guarantee he can politely send them back to the grocer; or if the case is an aggravated or flagrant one, turn the eggs over to the pure food officers for whatever action they may take in the matter. Usually, however, if the consumer deals with a reliable farmer he will very seldom have occasion to complain of the eggs not being as represented. Nor will there often be occasion for complaint against a reliable grocer who candles and guarantees his eggs.

Instead of a kerosene lamp, an electric light bulb may be hung inside the box.

The Fresh Egg.—The test of a fresh egg is its transparency and the smallness of its air space. There is no air space in an egg just as it is deposited warm and moist in the nest. As the egg cools, the contents contract and the air cell or air space appears. The shell being porous, the air space grows larger by evaporation all the time the egg is kept, the evaporation being fast or slow according as the temperature is high or low. Commercially an egg is fresh though it shows a small air space. If properly kept, an egg may taste perfectly fresh and pass in the market as a

A FRESH EGG
Note small air space.

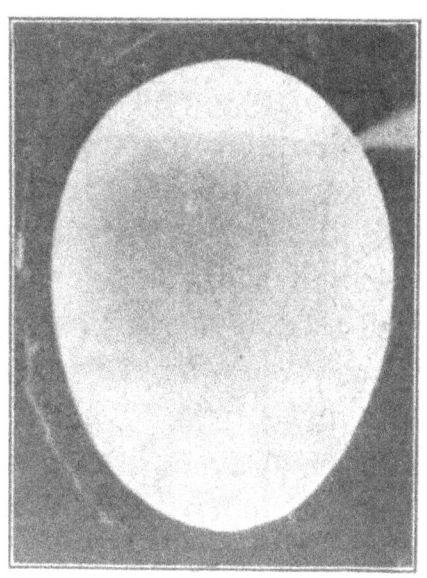

A STALE EGG
Note large air space, yolk settled to one side, showing dark.

fresh egg when a week old, or more. In some state laws it is enacted that when an egg has been kept or stored thirty days it is no longer a fresh egg. Kept under improper conditions, however, an egg would not test as a fresh egg, when two or three days old. The egg itself furnishes the evidence as to its freshness. The fresh egg is good legal tender at the country store, but when so used it should be treated as counterfeit currency by the time it reaches the consumer.

Marking Eggs.—The trade-mark of the egg producer is frequently stamped on the egg. This is easily and quickly done with a small rubber stamp one-half an inch in diameter or less. The name of the producer or the name of his farm, with address, may be stamped on the egg; also the words "guaranteed fresh." Where eggs are shipped through an association of producers the usual method is to give each farmer a stamp on which is the name of the association, with a number for each farmer. In this way if a consumer finds fault with the eggs he purchases he will send his complaint to the association, giving the number on the egg. The association manager knows from this number where the eggs came from, and the farmer is notified.

The stamp on the egg is the best advertisement the producer can have. If he can always furnish eggs of good quality it is worth his while to put his trade-mark on them. Every pound of butter or coffee, every 5-cent can of condensed milk, or loaf of bread, in fact about everything that the housekeeper buys from the grocer bears the stamp of the manufacturer. This advertises his goods and at the same time protects him against fraudulent imitation. If the poultryman wishes to build up and hold a trade for good quality eggs he should advertise by putting his stamp on them.

Where shipments are made direct by express or parcel post to consumers this may not be necessary, but it might happen that the customer would mix his eggs with some others, and the blame, if any, if the eggs were bad, might come back to him.

The eggs are frequently packed by the poultryman in cartons holding a dozen, and his name and address are printed on the carton. The carton is sealed, and the eggs are guaranteed if the seal is not broken. In such case the eggs are not stamped.

Summary.—The loss in quality of eggs is due to:

1. Improper feeding of the hens.
2. Dirty nests and yards.

3. Cracked eggs.
4. Broody hens.
5. Stolen nests.
6. Irregularity in gathering.
7. Storing in a warm place.
8. Keeping too long before marketing.
9. Fertility, or keeping males with the flock.
10. Exposing near stoves or in the hot sunshine.

There is a financial loss due to:

1. A loss in quality.
2. A system of buying by the dozen or case-count without reference to quality.
3. Indirect methods of buying which add to the cost.
4. Express rates which discriminate against shipments in small quantities.
5. Lack of co-operative effort between producers on the one hand, and between producers and dealers and consumers on the other.

Conclusion.—Improvement in quality will come when the producer who has eggs of good quality to sell insists upon the purchaser paying him according to quality, and, on the other hand, when the purchaser establishes the inflexible rule of grading eggs and paying according to grade.

Improvement in financial returns will come with improvement in quality; improvement in transportation; extension of the refrigerator service; more direct marketing and with co-operation between producers and between producers and consumers.

REFRIGERATION OR COLD STORAGE OF EGGS

The invention of the method and the growth of the business of preserving eggs by refrigeration has been one of the notable industrial developments of the United States in recent years. Whatever may be the merits of the cold storage product, the business must be recognized as one of great importance not only as it affects the cost of living or the food supply but as it affects the business and the profits of the poultry producer. Investigations have shown that there was put into cold storage during the year ending April 1, 1911, the enormous total of about 10,000,000 cases of eggs of an estimated value of over $64,000,000. Of this, practically 80% is handled in the three months of April, May and June, the percentage of the total being 42% for April, 25% for May and 12.5% for June. The cold storage of butter in the same period amounted to $40,000,000. It was reported that there were 500 cold storage plants in different sections of the United States storing eggs in that year.

It cannot be definitely stated just how much the poultry producer is the gainer or loser by the invention and development of the modern system of cold storage. There are those who claim that the business is an injury to the poultry industry, but the public has come to accept it as necessary for the proper distribution of food stuffs. In the early days of cold storage of eggs, and even in later days to a small extent, there was abundant excuse, for the violent antagonism that the business frequently encountered. It is true that cold storage eggs have been frequently sold as fresh, and even to this day in certain states unscrupulous dealers practice this fraud upon both the poultryman and the consumer in the absence of a law that would send them to jail for the act. But there is little excuse now for antagonism to the business of cold storage, though unscrupulous dealers and vendors of eggs may occasionally take advantage of it to enrich themselves at the expense of the public. The business of refrigeration has been more perfected and the quality of the product improved. Again laws have been enacted making it a criminal offense to sell storage eggs as fresh eggs.

The substance of the state laws that are now in force are that keeping eggs or other products in cold storage for thirty days makes them storage products. Any such product offered for sale must be stamped as such either on the package or product itself. The time limit of storage is, in New York, ten months; in some states nine months.

Effect on Prices.—With storage eggs thus stamped and sold, the poultryman can have no valid reason for objecting to the product. It is doubtful if the high price of selected fresh eggs will be appreciably affected by the sale of storage eggs. The people who buy storage or second grade eggs are not the ones who make the price for select eggs. It is pointed out that twenty years ago eggs sold during the surplus season in some states at 6 and 8 cents a dozen. Now they barely touch 15 cents at the lowest. There is no evidence, however, that this advance is wholly or in part due to cold storage, nor can it be proven that the much higher prices now received during the period of scarcity is due to cold storage. The fact is, however, that during the growth of refrigeration the price of eggs has been climbing upward.

There may be reason for the claim that the higher prevailing prices during the surplus season are due to the taking from the market of a large proportion of the eggs and putting them into cold storage. It is reported in the evidence of a Senate Investigating Committee that the daily consumption of eggs in New York City during the spring of 1910 was 12,000 cases. The receipts were about three times as much. What would happen to the market with receipts three times the consumption without a storage outlet? Clearly, the only thing that would save the eggs from being dumped into the harbor or thrown back on the farms as fertilizer would be such a reduction in prices that the people would consume the eggs. The storage business, therefore, tends to raise prices during the season of heavy production, but the tendency is in the opposite direction in the season when there is a deficiency in the supply.

The Refrigerator Egg.—The keeping of eggs at a steady low temperature is the most successful known method of preserving eggs. The best temperature for cold storage is 29 to 30 degrees. The principle of cold storage is that bacterial action, which causes decomposition or deterioration of the egg, does not take place at this temperature. The colder the eggs are kept without freezing the better. Fresh eggs of good quality may be kept at a temperature of 28 degrees, while those that are not perfectly fresh require a temperature of 30 to 32 degrees for best results. A steady temperature with a free circulation of air is absolutely necessary in the storage room. The eggs are stored in clean, odorless crates holding 30 dozen.

It is not possible to detect by candling any difference between a storage egg and a fresh one. There is a slight evaporation of the contents of the stored egg, but a fresh egg that has been laid for several days may show the same amount of air space. In a case lot of eggs as it is taken from storage, evidence of storage may be found in a slight mold which will show in cracked eggs.

The success of storage depends very largely upon the quality of the egg when it goes into storage. Storage does not absolutely prevent deterioration of the egg, it checks it. Under the best of conditions it is not as good as a fresh egg, but under proper conditions storage eggs are better than a great many fresh eggs that go to market during the warm months. An egg may technically be fresh and yet not be a good egg. Storage men have learned by dear experience that the early spring egg, the April egg in most sections, is the best for storage purposes.

Eggs laid in March, April and May and stored then are

in better condition than eggs laid in July or August when taken out of storage in November or December. During the warm months deterioration has set in before the eggs reach the refrigerator, and such eggs lose more in quality in the short time they are in storage than early spring eggs. This may not be due to a difference in quality of eggs when laid, but to the higher temperature to which they are sub-

CANS OF FROZEN EGGS

These cans hold 30 pounds each of separated whites and yolks, or whole eggs. Delivered to baker or confectioner in frozen condition. (Courtesy, Bureau of Chemistry, U. S. Department of Agriculture.)

jected before they reach the refrigerator. It is a question of handling the egg rather than a difference in the quality.

Limitations of Cold Storage.—The business is practically confined to large corporations with ample capital, located in large cities. Mechanical or artificial refrigeration is used, though ice plants are also used in a limited way. On account of the rapid deterioration that takes place after the eggs are removed from storage, it is an advantage to have the storage houses located near a large distributing center so there may be no unnecessary delay in getting the eggs to the consumers. It is doubtful if the business could be as successfully handled with smaller plants located near the points of production rather than centers of distribution, but in certain producing sections where there is a considerable local market small co-operative plants might be established with profit.

The Future.—Will the business of storage increase? With continued improvement in refrigeration and in extension of the service so that the egg will be better taken care of after it leaves the refrigerator and till it gets to the consumer, there will be a strong tendency to an increase in the business. Another factor, however, will be operating in the other direction. The producer by breeding better layers, fowls that will lay a larger percentage of eggs in the fall and winter, will be doing his best to put the cold storage plant out of business. But that is not imminent, desirable as it might be from the standpoint of the consumer as well as the producer. Greater progress must be made than has ever been made in poultry breeding if any one now living is to see the day when winter egg production is to equal that of the spring and summer. The best we can hope is that the poultryman will produce better winter layers, and that the winter layers will, in part, relieve cold storage of the burden of maintaining a proper distribution of eggs throughout the year.

Liquid Preservation of Eggs.—For home purposes eggs may be successfully preserved in a liquid preservative. Liquid preservation was formerly used commercially to a considerable extent, but the business has been largely superseded by the cold storage method. Where cold storage is out of the question a great many eggs are "put down" in some liquid preservative.

The Water-Glass Method.—A solution of water glass (sodium silicate) is the most generally used preservative for home purposes. Water-glass liquid or syrup may be obtained at most drug stores. The price is about 75 cents per gallon. It varies somewhat in quality. Thatcher, of the Washington Station, states that it should contain approximately one part sodium oxide to every $2\frac{7}{8}$ parts silicon oxide, and be of a consistency of about 38 degrees Baume.

It has been found that the best strength to use is about one part water-glass to 10 parts water. The water should be boiled and to every 10 quarts water add one quart water-glass, or in that proportion. The water must be allowed to cool before putting the eggs in. The receptacles used should be wooden buckets or kegs, or earthenware jars or crocks. Galvanized iron buckets or tubs may be used. Fruit jars may also be used. Metal vessels that will corrode in water should not be used. The liquid must cover the eggs, and then a little more, to allow for evaporation so that the eggs will always be covered.

Approximately three dozen eggs will fill a gallon jar, or ten times that number in ten gallons. It will require about four pints of the liquid to the gallon of eggs. The eggs should be kept in as cool a place as possible. The coolest part of the cellar should be used. The fresher the eggs are when preserved the better, but they may be kept a few days in a cool place before preserving. No cracked or thin-shelled eggs should be used.

Eggs preserved by this method will keep from the season of lowest prices to season of highest prices and be in condition to be used. They will not have the taste of the fresh egg, however. The white is thinner than in the fresh egg, but they will be perfectly wholesome. The water-glass closes up the pores of the shell, and in boiling the shell will crack. A puncture with a needle in the large end will prevent this.

The poultryman should understand, however, that no matter by what method they may be put down, preserved eggs are not as good as fresh eggs, and they should not be sold as such. Laws affecting the sale of cold storage eggs as fresh should apply equally to preserved eggs.

Selling Eggs for Hatching.—Improvement in breeds of poultry rests largely upon the work of breeders who sell eggs for hatching. The facility with which eggs may be shipped great distances and the comparatively small cost of shipping make it possible to secure good stock from successful breeders in any part of the country. The introduction of new and better blood is accomplished more often by the purchase of a setting of eggs than in any other way. At slight expense for express or parcel post it is possible to secure the best blood from the next county or from across the continent. This is one of the factors that make for the rapid upbuilding of the poultry industry. It is, of course, true that distance too often lends enchantment and that the farmer or poultryman could often secure as good blood from his next-door neighbor than from a distant state.

The facilities afforded by Uncle Sam and the express companies for securing new blood has been taken advantage of naturally by unscrupulous men who conduct a profitable long-distance business with the help of printers' ink and advertising. The very facilities for building up an industry are made the means for tearing it down. However, the good over-balances the evil, and it is the few who suffer. There is no remedy for the evil except that the purchaser learn to use ordinary business judgment in making his purchases and inquire into the reliability of the breeder before he sends him money, sending money to no one that he knows nothing about except what is stated in his advertisement.

Advertising, however, is a great factor in the distribution of improved strains or breeds of poultry. The business of the breeder is built up largely through advertising, and in proportion as he exercises judgment and skill in his advertising in proportion will he reap the financial reward of his success as a breeder.

Selecting Hatching Eggs.—The importance of exercising extreme care in the selection and handling of eggs for market has been emphasized. Greater importance, if possible, should be attached to the selection of eggs for hatching. The breeder who is doing an honest business will carefully cull the eggs before shipping. Only those of normal size and shape should be used. If the poultry breeders would make it a universal rule to set or sell no eggs for hatching that did not weigh two ounces each it would soon result in a vast improvement in the eggs of the country. That is one thing the breeder can easily control, selection for size of egg. Eggs should be clean, and preferably not washed. Washing injures their hatching quality, especially when shipped great distances. They should not be more than a week old before shipping, and kept in a clean, dry, cool place in the meantime. Further discussion of selecting eggs for hatching will be found under chapter on Incubation.

Packing Eggs for Hatching.—The result in hatching eggs shipped long distances will depend very materially on the method used in packing them. The packing must prevent breaking and jarring the eggs as much as possible. The package should not be air-tight, otherwise the eggs will sweat if subjected to wide ranges of temperature. Dry, clean excelsior, or wood wool, chaff or fine hay make good packing material. The eggs should not be wrapped in paper. Probably the most satisfactory shipping package is the split basket with handle. The basket with the lid is the most convenient. A muslin cover is sewed on to the basket that has no lid. The handles make the package convenient for lifting and at the same time prevent placing other boxes on top. There are other satisfactory shipping boxes. They should have some spring or resiliency to prevent undue jarring of the eggs. Each egg should be wrapped in excelsior or some other good packing material.

MARKETING POULTRY

At the present time the largest proportion of farm poultry is sold alive. The killing is done by the dealers in the city. It is done by them more cheaply and better, as a rule, than it can be done on the farm. The farmer and his help have not usually the skill to do the work properly. Another advantage of shipping the fowls alive is that the dealers in the cities, being in close touch with the demands of the market, can even up the supply to meet the demand. In the case of a surplus coming in one day, they can hold part of it over for several days and kill only sufficient to meet the immediate needs. If the poultry all came to market dressed there would frequently be a glut; that would mean often putting considerable quantities into cold storage or losing it. On the other hand, there would frequently be a dearth of fresh-killed stock. This would compel handlers of poultry to provide large storage facilities, and the consumers would be using storage stock a large part of the time instead of fresh stock.

The evil of this system of live shipments is that in most cases chickens are paid for on the basis of weight without regard to quality, though one farmer may furnish chickens with 25% more edible meat for the money received than another farmer. While the great bulk of the poultry is shipped alive, special markets may be worked up by farmers or poultrymen for dressed chicken of special quality by shipping direct to consumers. That is the best way and about the only way to get full value for fowls of good quality.

Large quantities of live poultry are shipped both east and west from the Mississippi Valley states and to a limited extent from other sections. Many carloads of such poultry

POULTRY HOUSE EXHIBIT
Part of an Oregon poultry demonstration train.

are shipped from the central west to Pacific Coast points and as far east as New York. Special live poultry transportation cars holding from 4,000 to 5,000 fowls are used. A man accompanies the car, doing the feeding and watering from an aisle in the center of the car. A rental is charged for the cars in addition to the freight. Another important development of market methods is the purchasing by the meat packers of large quantities of farm poultry for fattening. A large proportion of this poultry, after being fattened, is killed and put into cold storage to hold for a rise in price.

Killing and Picking Fowls.—Success in marketing dressed chickens to a select trade depends very largely on methods followed in killing and dressing. It should be remembered, always, that cleanliness and neatness have a market value when applied to dressing fowls and packing

UNLOADING A NEBRASKA CARLOAD OF POULTRY AT SAN FRANCISCO

them for market. In other words, the basis of a select trade must be superiority of goods.

Killing.—Before being killed the fowls should be starved 24 hours in order to empty the crop. This will make the fowl more attractive. By starving, the intestines will be largely emptied of their contents. This improves the keeping qualities. It is important in killing that the fowl be thoroughly bled. A well bled carcass looks better and keeps better. The dressed fowl will not be select where the bleeding has not been done thoroughly.

Sticking.—The best method of bleeding is that of sticking the fowl in the mouth. It takes some practice to become expert. The success in bleeding depends on the kind of stick. Usually the bird is hung up on a level with the shoulders of the operator. The head of the chicken is laid in the left hand so as to have the comb down. The pressure should be on the boney part of the head, not on the neck, as pressure there will prevent proper bleeding. The bills are held apart by inserting the first finger in the corner of the mouth. The knife should be sharp pointed, about 2 inches long and a fourth of an inch wide. The blood vessel is first cut on the right side of the roof of the mouth at the neck where the bone of the skull ends.

The brain stick is necessary where the fowls are to be dry picked. This is to "loosen up" the feathers. After cutting the artery the knife is quickly inserted in the brain through the groove in the roof of the mouth. This paralyzes the muscles and makes dry picking easy, but it must be done before the muscles contract. The picking commences immediately after the brain stick is made and even before the fowl dies.

Picking.—The breast feathers are first picked, then the long tail feathers and wing feathers. The picking must be done quickly when the feathers come off easily. Care, of course, must be taken not to tear the skin. After the rough picking comes the pin feathering. This is done sitting, with the chicken on the knees. It is necessary to use a knife to catch all the small pin feathers.

This method of killing and picking requires considerable practice to be able to do it at a profit. In commercial establishments expert pickers kill and dress in this way as many as 100 fowls per day.

Cooling.—When picked they should be immediately put into cold water for about an hour. This removes the animal heat and improves their keeping quality. It also gives them a plumper appearance. They should never be packed until the animal heat has been removed.

Shaping.—After picking and cooling the bird is frequently shaped on a shaping board to give it a more compact appearance. It is placed breast down on a board and a weight put on its back. This may be called a harmless "trick of the trade." Fowls that are naturally well fleshed and plump will not be improved by this treatment.

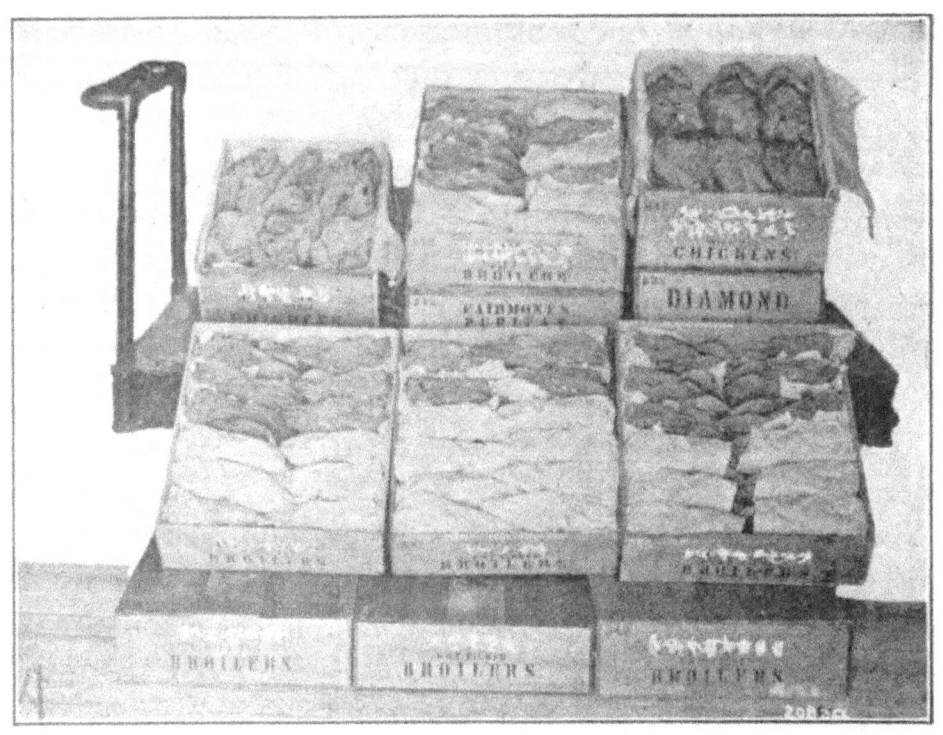

DRY PICKING, DRY COOLING AND DRY-PACKED POULTRY

Scalding.—For home use and immediate consumption scalding is the almost universal method. It is also used by dealers in many large centers. It is the easiest and quickest method of removing the feathers. The objections to scalding are that it tends to disfigure the skin and change its natural color. Water for scalding should be kept just below boiling. The feathers rub off easily when properly scalded. The legs and feet should not be allowed to touch the water.

Drawn and Undrawn Poultry.—It is a much debated point as to whether fowls should be drawn when killed. In some markets the law requires poultry to be drawn before exposing for sale. In most markets, however, the practice is general to leave the drawing or dressing to the retail dealer at the option of the purchaser. The evidence seems to favor the view that undrawn poultry keeps better than drawn. The theory is that in the drawn chicken the inside of the intestinal walls are exposed to the invasion of bacteria which will hasten the process of decay. On the other hand, it is claimed that the putrefactive bacteria of the intestines will infect the flesh of the fowl and cause more rapid decomposition. Further investigations seem needed.

Meantime there is no ground for wholesale condemnation of undrawn poultry. With proper bleeding and chilling no danger may be feared from either drawn or undrawn fowls. One important point in favor of the undrawn is that the housekeeper could see evidence of unfitness for eating if she drew the fowl or removed the viscera herself. Most of the important diseases of fowls are often indicated by the appearance of the liver and intestines. In the case of tuberculosis the evidence, in a great majority of cases, is found in the condition of those organs.

Parcel Post Shipments.—While the practicability of parcel post for dressed poultry has not been demonstrated in an extensive way, it affords a medium not heretofore available for direct shipments to consumers. For a special trade in fancy stock it offers an opportunity to the poultryman who can produce the proper grade of stock to do a profitable business. Dressed chickens can be taken direct from the farm to the door of the consumer in the city within 150 miles at a cost of from 1 to 2 cents a pound, depending on the weight of chicken in package. The great waste of marketing should be saved to both the consumer and to the producer by direct shipments. This should also encourage a much greater consumption of poultry.

Loss in Killing and Dressing.—This loss is usually considered the weight of feathers plucked and the blood drawn from the fowl. The head and feet are left on and the carcass undrawn. The following table shows results secured at the Oregon Station from 88 fowls:

Liveweight	348	pounds
Dressed weight	306.8	"
Loss	41.2	"
Per cent. loss	11.8	

The **dressing percentage** varies with the condition, age and breed of fowl. A fair average would be 88%.

Edible Meat on the Fowl.—The percentage of edible meat varies with the breed, the condition of the fowl and its age. Work at the Oregon Station indicated that a fowl in fair condition has about 60% edible meat. The waste was found to be about as follows for a fowl weighing seven pounds:

Feathers and blood	8	ounces
Offal	27	"
Head, bones, shanks	8.2	"
Total loss	2 lbs. 11.2	"
Edible meat	4 lbs. 4.8	"

Capons and Caponizing.—Capons are castrated males, or males with the sexual organs removed. The operation of removing the testicles is called caponizing. These organs are within the body cavity of the fowl attached to the back and lie close to the lungs and heart. The operation is a delicate one, and special instruments are made for the purpose. By following directions closely the amateur may soon become expert. Full instructions for operating are furnished with the instruments, and these need not be repeated here. The operation consists in making an incision near the thigh and between the two last ribs and removing the testicles with the proper instrument.

The object in caponizing is to produce a better quality of flesh and to make the surplus cockerels more marketable. A capon will sell for practically twice as much per pound, and often more, than a mature rooster. Under present conditions, however, it will pay to sell the males as broilers when the broiler market is good. This applies where the chicks are hatched early. The capon market, except in isolated cases, is not yet highly developed. It will develop as consumers become educated to the superior meat quality of the capon. As indicating the possibilities, it may

DRESSED CAPON

HENRY DANA SMITH (MASS.)

Who caponizes 4,000 cockerels a year and sells them as "soft roosters." Mr. Smith is very expert and averages about 50 an hour. (Photo by A. G. Lunn.)

be stated that capons, canned in France, are for sale in the large cities of this country and at high prices. If cockerels are to be kept till the fall it will pay the farmer to caponize them and keep them till January or February, when the market is good for roasting chickens. Capons are quiet and docile, do not crow or fight, and sometimes make excellent mothers for chicks.

CHAPTER XVII

DISEASES AND PARASITES OF FOWLS

A knowledge of poultry diseases is of value to the poultry-keeper more in enabling him to locate unfavorable hygienic conditions than in the curing of diseases. In the discussion of poultry management in general the author has endeavored to keep prominently to the fore the great importance of proper sanitary conditions as a means of maintaining health or of avoiding diseases as much as possible. In other words, the poultryman must rely rather on preventive measures than on curative treatment to maintain his flock on a healthy, profitable basis. It is an unprofitable business to be continually fighting diseases and treating sick fowls when a knowledge of simple hygienic rules will enable the poultry-keeper to prevent diseases and obviate treatment. As a rule, it does not pay to treat sick fowls. An individual fowl, on the average, is worth too little to pay to treat; besides fowls suffering from contagious diseases are a menace to the rest of the flock and the sooner they are gotten rid of the better.

There are, however, certain diseases or ailments that are amenable to simple treatment, and if the poultryman possesses the requisite knowledge of the ailment and its treatment, he may often save himself considerable loss.

Hygienic Conditions.—The importance of the subject warrants recapitulation here of what has already been emphasized in different chapters of proper sanitary or hygienic conditions.

Fresh Air.—Fresh air is not only an egg producer but a health preserver as well. Many of our poultry diseases are the result of keeping the fowls in ill-ventilated houses. A lack of vigor is often the result of impure air. Diseases of the respiratory organs, such as catarrh, roup and colds, thrive only in ill-ventilated houses. It is useless to treat for such diseases fowls that are kept in houses that breed disease by bad ventilation.

Fresh Ground.—Next to fresh air, fresh ground is the best preventive of disease. Many diseases having to do with digestive organs thrive where no attention is paid to keeping the ground on which the chickens run fresh and clean. Tuberculosis, cholera and other diseases are usually contracted by the fowls picking up from the ground feed that has come in contact with the germs of the disease. Various parasites, such as gapes and tape worm, are taken up by the fowl in this way. An unclean feeding-ground is a fruitful source of disease.

Fresh air and fresh ground are the cheapest things at the command of the poultry-keeper and when he learns to make full use of them there will be comparatively little danger from poultry diseases.

No flock of chickens, however, is entirely immune from diseases. In spite of the best sanitary conditions, diseases will sometimes get into the flock and remedial measures will be necessary.

Cleanliness.—The nest boxes should not be a breeding place for germ diseases and insect pests. They should be frequently cleaned and disinfected. The droppings should not be allowed to accumulate, and on no account should the night droppings be allowed to fall and mix with the litter on the floor if the floor is used for a feeding- or scratching-ground. The litter should be kept reasonably clean and dry.

Disinfection.—The culture treatment of yards is discussed in another chapter. If cultivation and cropping can be regularly and thoroughly done there will be little need of other treatment of the soil to destroy infection. It may sometimes be necessary to disinfect the feeding-grounds. The most common method, probably, is the use of quick lime. Fresh lime should be air-slaked and broken into a floury powder and sprinkled over the ground until white. To prevent burning the chickens' feet the earth may be raked over it a little. For the houses various disinfectants are used. For bacterial diseases such as roup, tuberculosis, and cholera, formaldehyde may be used at the rate of 1 pint to 20 gallons of water. This should be applied with a spray pump. The walls, ceiling, floor, roosts and nests should be thoroughly drenched with the spray.

Boiling water may be used for articles such as drinking vessels, small feeding-troughs, etc., dipping them in the water. Commercial germicides such as Zenoleum and Kreso dip may also be used as disinfectants. In small houses that may be closed up tight, probably the most effective disinfection is to fumigate with formaldehyde gas. Use at the rate of 16 ounces of 40% formaldehyde to 6 ounces of permanganate of potash, per 1,000 cubic feet air space. Put the permanganate in a jar and pour the formaldehyde into it and then quickly leave the house and close the door. Leave the house closed for two or three hours. This is a convenient and effective method of disinfecting incubators as well as houses.

Disinfecting Drinking Water.—Germ diseases such as roup, canker and chicken-pox are frequently spread among the flock through the water in drinking vessels. Where there are any indications of such diseases it is well to use permanganate of potash in the drinking water, using about a fourth teaspoonful to a gallon of water. Permanganate ordinarily is cheap, and should be liberally used.

It is known that disease germs are more virulent after

they have passed from one animal to another. The germs may be passed in the excrement or droppings of the diseased fowl even before the symptoms of the disease are manifest. They may be passed in countless numbers, and if the droppings come in contact with the food another fowl takes them up and the germs, in all likelihood in a more virulent form, enter the body of another and cause disease.

Tuberculosis.—Tuberculosis is a germ disease and is probably the most destructive of all diseases of mature fowls. The disease progresses slowly and may be well advanced before the symptoms are noticed. Avian tuberculosis was first reported in this country by Professor Pernot of the Oregon Agricultural College in 1900 (Bulletin 64), and to him the author is indebted largely for the facts herein presented in regard to this disease. The most pronounced symptoms of tuberculosis are lameness and loss of flesh. Tubercular fowls, however, often have the disease without lameness, but lameness is often associated with the disease. On the other hand, lameness does not always indicate tuberculosis, as it may be due to other causes. In advanced stages there is great loss of flesh or wasting. There is no loss of appetite.

Seat of Disease.—As the organisms enter the body with food, the disease is more commonly found in the digestive tract and the liver than in any other part of the anatomy. Many cases of the disease in its advanced forms fail to show any lesions of the lungs. There are two common forms that are easily detected; one is a fibroid growth on the intestines varying in size from a pinhead to a lump as large as a walnut. In cutting through these tubercles, they will be found to contain a substance varying from a serous fluid to a rather dry, cheesy mass according to their size and age. It frequently happens that when a tubercle on the intestine becomes the size of a large pea, the mucous membrane and wall of the intestine on the inner side of the tubercle breaks down and discharges the contents of the tubercle into the fœcal matter that is passing through the intestine, thus carrying out with the excreta a great number of living tubercle bacilli.

"The liver is the other organ commonly affected. When the tubercle bacillus finds its way into the liver and begins to grow, a yellowish spot is soon formed, increasing in size as the disease progresses. The structure of the tissue at this point is changed to a hard granular mass containing within it the bacilli and the same substance as found in the intestinal tubercle. The growth of the tubercles necessarily increases the size of the liver until it sometimes becomes twice its normal size, and the tubercles are frequently so numerous as to give the liver the appearance of peanut taffy.

"There are other spots of similar appearance sometimes found on the liver that must not be mistaken for tubercles. A crude way of distinguishing tubercular lesions is by the fibroid tissue of a tubercle being tougher and harder than the structure of the other spots mentioned, and by the center being filled with a substance as before described. Sometimes the disease is scattered all through the internal organs, and tubercles may be found even on the heart."— *Pernot.*

The only certain method of diagnosing the disease is a bacteriological examination. The germ is a small organism measuring on an average 3/25000 of an inch in length and can only be seen through a microscope. Poultrymen should avail themselves of the services of the bacteriologist of the experiment station if they are suspicious of this disease in their flocks. Many of the stations have facilities for doing this work without charge.

It is not definitely known that bovine or human tuberculosis is transmitted to fowls. Pernot, though recognizing different types of tubercle bacilli, recognizes the possibility of transmission and urges caution. Other investigators have failed to produce the disease in the fowl with the bovine or human bacillus.

Tuberculosis is not transmitted through the egg to the chick. Some investigators point to the possibility of transmission, but the possibilities are so remote as to be without significance to the poultryman.

There is no known cure for the disease. Proper sanitation and prompt destruction of affected fowls must be relied upon to prevent the ravages of the disease. There is no reason for alarm if the poultryman makes full use of fresh air, fresh ground and sunshine in the management of his flock. If particular care be taken in this respect the disease will not get much headway. The frequent renewal of the stock, killing off the old and replacing them with young, is a favorable factor in the control of the disease. The poultryman would do well before purchasing fowls to inspect the flock from which they come and secure them only from flocks that show no indications of disease.

A Tuberculin Test.—Until the year 1914 there was no known method of testing live tuberculous fowls. Bovine tuberculin has been proved valueless for this purpose. In that year Dr. Van Es of the North Dakota Station discovered that avian tuberculin, when properly used, is an almost certain test of the disease. The tuberculin is injected into the comb or wattles. The injection must be made near the surface, but not so near that the fluid may burst through the epithelium. In the experiments noted the results were ascertained in from 24 to 72 hours. The reactions consist of a swelling and discoloration of the part injected, the size of the swelling varying considerably in different cases. The swellings or reactions, in Van Es's experiments, indicated a tuberculous condition in 88 cases out of 90. On the other hand, 8 to 9% of the fowls showing no reactions were found to be tuberculous.

Should this test prove to be as successful in the hands of others as it has been in the experiments reported, it is a discovery that marks a most important advance.

Roup.—Many poultrymen believe that roup and the kindred affection, catarrh, are the most troublesome diseases of poultry. Roup proper is believed to be a contagious germ disease. Catarrh, exhibiting practically the same symptoms, is not contagious, being produced usually by improper housing. The specific organism producing roup has not been discovered.

A BAD CASE OF ROUP IN ITS ADVANCED STAGE

(Courtesy, Prof. T. D. Beckwith, Bacteriological Department, Oregon Experiment Station.)

Symptoms.—It usually begins with a watery discharge from the nostrils and eyes, which as the disease progresses, becomes thicker and of the nature of pus. The nostrils become clogged, interfering with breathing, and there is usually a swelling around the eyes. The swelling often grows until the eye is closed entirely. The disease frequently spreads to the mouth and throat and assumes the character of diphtheria, when death soon results.

Treatment.—The only hope of curing is in recognizing the disease in its first stages and applying remedies. Permanganate of potash of a 2% solution has been successfully used. The head of the bird should be dipped in the solution and held there as long as possible without strangling the bird. The success of the treatment depends upon getting the solution into the nostrils. This treatment should be continued two or three times daily until a cure is affected. Kerosene is also an effective remedy. The face should be washed with a feather dipped in the oil and a little oil injected up the nostrils. If the mouth or throat are affected they should also be swabbed out with a feather dipped in the oil. Peroxide of hydrogen is also used successfully for injecting into the nostrils and swabbing out the throat. If the swelling on the face has reached a stage that pus has formed, an incision should be made, the pus removed, and the sore washed out with the permanganate solution or with the peroxide of hydrogen. When the disease has reached that stage, however, treatment will not often be successful and it will not pay unless the fowl has some special value. The sick fowls should be isolated and the premises disinfected. Care should be exercised in introducing new fowls, and it is a safe practice to put them in quarantine several days before putting them with the rest of the flock.

Catarrh.—Possibly in the large majority of cases, what is thought to be roup is simply catarrh or colds. The symptoms are practically the same. The treatment of affected fowls recommended for roup may be followed for catarrh. When colds or catarrh appear it is a sure indication that something is the matter with the housing. The fowls may be crowded too closely together on the roost; there may be cracks in the walls through which the wind blows strongly on the chickens, or there may be insufficient ventilation.

Diphtheria.—This is not an uncommon disease among fowls and it is very fatal. A false membrane grows in the mouth and extends down into the throat. Treatment is not often successful. Kaupp (Colorado, Bulletin 185) recommends the burning of the diphtheritic patches of the mouth with stick nitrate of silver (lunar caustic). A 2% solution of pure carbolic acid in water applied three times daily to affected birds is also recommended. When it can be done without causing bleeding, the diphtheritic membrane should be removed and the application of carbolic acid continued. The germ of fowl diphtheria resembles that of the human species. Attempts at the Oregon Station to reproduce the disease in chickens by human baccili failed, though further investigation seems necessary to settle this point. In the meantime, poultrymen should exercise care in the handling of fowls affected with this disease.

Chicken Pox.—This is a contagious disease and most prevalent in damp weather. Small crusts or wart-like spots appear, sometimes on the face, sometimes under the wings and on different parts of the body. A simple and effective remedy is to apply to the birds affected carbolated vaseline or sulphur ointment.

Cholera.—This is the most fatal of all diseases, though not as general as roup and tuberculosis. It is comparatively rare. The symptoms are diarrhœa, loss of appetite, excessive thirst, pale comb and wattles and extreme exhaustion. Death occurs in from a few hours to two or three days. There is no cure. Vigorous measures of disinfection must be taken.

Canker.—Canker is indicated by white or yellowish spots in the mouth and throat and corners of the mouth. Peroxide of hydrogen is effective. Powdered chlorate of potash blown through a glass tube or straw onto the spots is also recommended. Use permanganate of potash in the drinking water.

Diarrhœa.—There are various causes for diarrhœa besides those already mentioned. It may be caused by improper feeding, chilling, filthy drinking water, decayed meat, and irritating matter in the intestines. A tablespoonful of olive oil or 25 grains of epsom salts per fowl, dissolved in water, is recommended. Boiled rice and boiled milk are also effective. Dry middlings are also beneficial in certain cases. Decreasing the quantity of laxative foods such as bran and wet mashes may often be all that is necessary.

Dropsy.—Abdominal dropsy is indicated by a heavy hanging abdomen. The abdomen feels soft and watery. It is due to a rupture of the blood vessel which permits the water to escape into the abdominal cavity. Treatment is not profitable, though temporary relief may be given by puncturing with a needle, or milk tube, which will permit the water to escape. Dropsy sometimes indicates a tuberculous condition.

Bronchitis.—Bronchitis is caused usually by draughts in the poultry house. It is found associated with catarrhal roup and is indicated by coughing and rattling in the throat. An effective treatment is two or three drops of spirits of camphor in a teaspoonful of glycerine, two days in succession. Two grains of black antimony in the food is also recommended. Swab the throat with permanganate of potash.

Peritonitis.—This is an inflammation of the peritoneum or membrane that covers the abdominal cavity. Successful treatment is difficult. Three or four grains of tincture of aconite in half a glass of water, giving a teaspoonful three or four times a day, is recommended. Frequent application of moist flannel cloths is beneficial.

Rheumatism.—Poultry kept on damp ground or in damp houses with restricted exercise, are subject to rheumatism. The fowl in walking has a jerky gait. Lameness does not always signify rheumatism. The limbs should be bathed in warm water or the fowls made to stand in warm water, then rubbed dry and a mixture of turpentine and sweet oil or camphor oil applied.

Apoplexy.—Apoplexy is due to the bursting of a blood vessel of the brain. Treatment is impossible, as the fowl usually dies very suddenly without indicating the disease. A fatty condition is usually the cause. Less starchy foods should be fed and more exercise given.

Limber Neck.—Apparent paralysis of the neck muscles is the symptom of this disease. The neck is limp and stretched out in front of the bird with the beak usually touching the ground. It is due to impaction or stoppage of the stomach. A tablespoonful of olive oil or castor oil will usually effect a cure.

Wry-Neck.—In this case the fowl has also apparently lost control of the neck. Instead of the neck being stretched out in a horizontal direction, the head is drawn back and down toward the body, the bird twisting it from one side to the other. This disease is usually associated with an over-fat condition, produced by a lack of exercise and feeding heavily on fat-producing foods. Epsom salts should be given, the ration changed and more exercise furnished.

Crop Bound.—This is indicated by a full and extended crop which is rather hard to touch. Foods of a fibrous nature or indigestible articles such as long, tough grass, which have been greedily eaten, produce crop bound or crop impaction. Irregular feeding may cause the fowl to over-eat at one time and produce the trouble. The materials in the crop become so wrapped together and impacted that the passage to the stomach becomes obstructed and the fowl gets no nourishment. Hunger increases, the fowl eats more, and the ball of food in the crop becomes larger and larger. The grain foods swell, causing further distention. The fowl finally dies of starvation with an over full crop. Simple treatment, however, will save the life of the fowl.

It is sometimes sufficient to give about a tablespoonful of olive oil to soften the mass, then with the fingers manipulate the mass until it becomes soft and moist. It may require an hour to do it. If the mass does not break up try holding the bird by the legs, head down, and gently work the food out of the mouth. If this is unsuccessful, resort must be had to an operation. After removing the feathers, an incision 1½ inches long should be made in the outer skin of the crop, then a small opening into the crop. With a small spoon or pair of tweazers, or the fingers, the contents may be removed, after which the lining of the crop and the outer skin should be carefully sewed together, separately. The wound should then be rubbed with vaseline. Feed the fowl lightly for a few days with easily digested food.

Chick Mortality.—Poultry raisers sustain great losses in the rearing of chicks. The losses have been so great in many cases as to drive the poultry raiser out of business. A large part of the loss is ascribed to what is called white diarrhœa. It should be clearly understood that there are different forms of diarrhœa in chicks. Diarrhœa may be caused in many ways. Possibly in a great majority of cases where the losses are heavy the diarrhœa in brooder chicks is due to chilling. Improper feeding will also cause diarrhœa. In such cases the loss is not due to an infectious disease over which the poultryman has no control. Again, large losses of chicks occur in the brooder from apparently no other cause than a lack of vitality.

Wrong methods of incubation, which are discussed in the chapter on hatching chickens, are often the direct cause. Lack of vigor in the breeding stock is often the cause of low vitality in the chicks. A hen failing to sit properly and contentedly on the nest will hatch chicks that show lack of vigor. Incubators that have not held the temperature steady or have not supplied the proper moisture conditions, will hatch chicks of low vitality. Such chicks are susceptible to bacterial and other diseases that would not affect strong, vigorous chicks. Small chicks are always very susceptible to environmental conditions, and where these conditions are found to be unfavorable it is quickly evident in the death-rate of the chicks.

NORMAL HEN'S OVARY
(Courtesy, Storrs Experiment Station.)

White Diarrhœa (*Bacterium pullorum*).—From recent investigations, it is clear that bacteria are responsible for a large part of the chick mortality. A certain germ, *bacterium pullorum*, was isolated at the Storrs Station which proved to be the direct cause of what is popularly known as white diarrhœa. Diarrhœa is but a symptom of the disease, which should not be confounded with various other kinds of diarrhœa. The germ was found in the fresh

DISEASED OVARY
Showing white diarrhœa condition. (Courtesy, Storrs Experiment Station.)

egg and in the ovaries of the hen, as well as in the chick when hatched. A diseased ovary produced a diseased ovum or egg, and a diseased egg produced a diseased chick, and a diseased chick may infest many other chicks in the brooder. It was also proved that the infection may be carried from the adult hen to another through the medium of the feed. Chicks of low vitality succumb more readily to the infection than those of good vigor. Again, chicks hatched in winter and late fall are not so subject to the disease as those in late spring and summer.

The influence of vitality is very clearly apparent. How far we can count upon vitality to ward off the disease or to maintain immunity, is not clearly established by the experiments. They emphasize the importance, however, of maintaining at all hazards the vitality of the stock. They also offer a possible explanation of the usually larger death-rate of chicks in large flocks than in small ones. One hen's chicks may be affected, another's may not. If the chicks from two hens are brooded separately, the chances are the one lot will live and the other may die. If they are brooded together, they all may become affected and all may die.

In white diarrhœa the deaths usually occur when the chicks are under four weeks of age. In describing the symptoms of the disease, Woods says: "The weakling is almost always big bellied, the abdomen protruding to the rear so that it punches out behind, and out of line with the vent, with the result that the chick looks as if the tail-piece and backbone had been pushed forward and in just above the vent." Upon dissecting the chick the following conditions will be found:

"*Crop.*—Empty or partially filled with slimy fluid or with food.

"*Lungs.*—Apparently normal. (Tubercles not observed.)

"*Liver.*—Pale, with streaks and patches of red. The congested areas are usually large in size. Occasionally epidemics will be met with in which the liver is more or less congested throughout. In such cases the portion of the stomach lying in contact with the liver is inflamed.

"*Kidney and Spleen.*—Apparently normal.

"*Intestines.*—Pale, and for the greater part empty. A

TWO WHITE DIARRHŒA CHICKS
Showing characteristic dumpy appearance.

small amount of dark grayish or brownish matter frequently present.

Ceca.—With few exceptions but partly filled with a grayish soft material. Only occasionally cheesy or firm contents.

"*Unabsorbed Yolk.*—Usually present, varying in size from a pea to a full-sized yolk. The color may vary from yellow to brownish green or nearly black. In consistency there is also much variation. It may appear perfectly normal, distinctly gelatinous, or watery. Frequently it looks like custard and again it is more or less dry and firm. Unless the chick has been dead for some time the yolk is not putrid, but merely stale.

"The chick as a whole appears more or less anæmic and emaciated. The muscles of the wings, breast and legs may be almost completely wasted away." (Bulletin 74, Storrs Station.)

The remedy suggested is the use of sour milk, though this is rather in the nature of a preventive than a cure. The chicks usually become affected before they are four days of age and very seldom after that. It has been found that by feeding sour milk just as soon as the chicks are ready to eat or drink, the ravages of the disease may be checked. Whether the lactic acid germ of the sour milk kills the white diarrhœa germ or whether from the sour milk the chick derives the strength and vigor that enables it to throw off the disease, has not been very clearly shown. At any rate, the Storrs experiment offers strong endorsement of the practice of feeding sour milk or buttermilk to young chicks.

That there are other disease germs which prey upon the young chick has been demonstrated at the Oregon Station. A different organism was found in chicks dead in the shell and in hatched chicks that died later with symptoms of white diarrhœa. When healthy chicks were inoculated with the germ it proved fatal, though when healthy chicks were brooded in the same brooder as the others they were apparently unaffected.

So far as the white diarrhœa investigations have gone it has been established beyond doubt that it is a bacterial disease. No remedy has been discovered. It has not been shown, however, that the poultry-keeper is helpless before its ravages. The encouraging feature of the situation is that high vitality in the chicks seems to carry a certain immunity or power of resistance to the disease, and until further light is thrown upon the subject, it is just as well to accept the theory that it is a disease, which if not the result of low vitality, need not be greatly feared where the health and vigor of the stock is unquestioned.

The Agglutination Test.—While there is no known cure for bacillary white diarrhœa, recent investigations by Jones of Cornell University have indicated an accurate test for

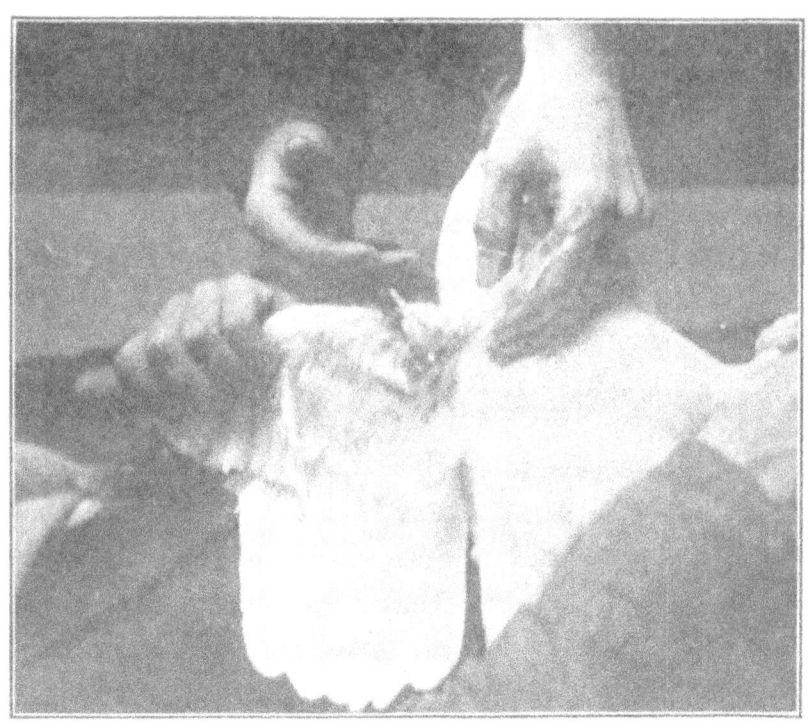

TAKING A BLOOD SAMPLE FOR WHITE DIARRHŒA TEST
(Photo by C. S. Brewster.)

infected fowls. It is called the agglutination test. By this test it is possible to determine whether or not adult fowls are infected. The importance of the test lies in the fact that it is possible for the poultryman to eliminate this disease in chicks by breeding only from fowls that the test shows are free from it. Several of the experiment stations have facilities for making these tests for poultry breeders.

An explanation of the test is given by Dr. Gage of Massachusetts as follows:[1]

"The two important biological factors necessary for making the microscopic agglutination test are (1) a test fluid containing a suspension of the organism causing the disease, and (2) a sample of blood serum from the individual to be tested, and the test is based on the fact that the blood sera of infected and non-infected birds when mixed with the test fluid react differently. The serum of the former, because of the presence of an agglutinin, a substance formed in the body of the bird because of infection with *Bacterium pullorum,* is capable of producing, when brought in contact with a suspension of the organism, a clumping together of the bacteria, a phenomenon which blood from non-infected birds does not show."

PARASITES OF FOWLS

The poultry-keeper must be able to cope with parasitic enemies or they will put him out of business. If every living thing has its own particular pest, the fowl has its full share, probably more than its share. There are a dozen or two insect pests or parasites that have no other business in life, apparently, than that of making life a burden to the chicken. We do not know how many. The number of varieties, however, is of no consequence compared with the number of individuals of any one variety that may be propagated or born into the world in a few days. From one single louse in the third generation, there may be produced in eight or nine weeks over 100,000 individual lice, each one hatched from an egg.

The different varieties work in different ways. One variety sucks the blood from the chicken and when tens

[1] Massachusetts Bulletin No. 163.

of thousands of these bloodthirsty villains are plying their trade, the hen will soon be pumped dry of blood. Others do not suck the blood but irritate the fowl beyond endurance by moving about or running foot races, possibly with 10,000 other entrants, on the skin. Others burrow into the skin or flesh; others suck the liquid contents from the cells of the skin and exude a poison under the skin. Still others do their damage by carrying infectious diseases from one fowl to another.

The internal parasites affect the wind-pipe, the stomach and intestines and cause various derangements.

Poultry parasites are divided, therefore, into two classes, external and internal.

External parasites may be divided into two kinds, namely, mites and insects.

Mites.—The chicken mite (*Dermanyssus gallinea*) causes more loss to the poultryman than any other species of mite and probably more than any other kind of insect or louse. These mites breed on the under side of the roost porches, especially where there is a rough surface and small cracks or crevices. They also breed in the cracks of the walls near the perches and in the nest boxes. Their presence will often be indicated by white dust-like patches on the walls. They are not found in any numbers on the fowl during the day but they crowd out of their hiding-places onto the fowls at night and suck the blood, then go back to their hiding. During the warm days of spring and summer they multiply rapidly.

Frequently sitting hens die on the nest, being literally bled to death by the pests. Sometimes they multiply so rapidly that they can be gathered by handfuls in nests or other places where they are undisturbed, especially under sitting hens. They live several weeks after being filled with blood. Under certain conditions they have been known to live several months. In size the mite is about 1/40 of an inch in length. If placed side by side 100 mites will cover a space of one square inch.

Control.—The mite, though possibly the most destructive of any poultry parasite, may be easily controlled. Various control methods are used.

Treatment 1.—Kerosene, crude petroleum, and distillate are effective. These will kill any mites they come in contact with. The oil, however, may not destroy the eggs of the mite. If the house is badly infested the whole interior should be thoroughly sprayed. In a week or ten days the application should be repeated to kill those that may have hatched after first spraying, and if necessary a third spraying should be given. If this is thoroughly done the mites may afterward be controlled by spraying the perches with kerosene or distillate, or a brush may be used and the oil applied all around the roost. The nest boxes should also be treated.

Treatment 2.—Instead of using coal oil or distillate in spraying the roosts, carbolineum or other tar preparations may be effectively used. Carbolineum is more effective than kerosene for the reason that it will destroy the mites' eggs as well as the mites when it comes in contact with them. Nests of sitting hens should be thoroughly painted before sitting; also the brooding coops. With any reappearance of mites the application should be repeated. The paint should be dry before the hens are allowed to use the nests or roosts. It will soil the feathers and may affect the flavor of the eggs. Brood coops should be thoroughly dried after painting or the chicks may be injured. Crude carbolic acid and kerosene or distillate, one part of the former to three of the latter, is very effective for mites, applied as a paint on the roosts and nests.

Treatment 3.—To five gallons whitewash add one pint

crude carbolic acid. Spray as above or use whitewash brush for applying it.

Treatment 4.—Where lime sulphur spray is used for fruit trees it may also be used for spraying the poultry house. If thoroughly done this should control the mites.

Lice.—Lice are not so injurious as the mites but they must not be allowed to breed unchecked. Unlike the mites they do not suck the blood but subsist upon the productions of the skin and the feathers. They live and breed on their host. There are three kinds of lice generally recognized. First, those that are found on the head and neck of the fowl and especially on young chicks. The scientific name is *Goniodes eynsfordii*. Second, the wandering lice (*Menopon pallidum*). These are found on different parts of the body. Third, those found between the barbs of the wing and tail feathers (*Liperurus variabilis*).

The conditions which encourage the breeding of these lice are filth, dampness and darkness in the poultry house. The eggs are laid among the feathers and attached to them, especially to down feathers. They hatch out in from six to ten days, the time varying. Lice will live several months without the hen or host. Theobald reports keeping *Menopon pallidum* for nine months on fresh feathers, they apparently eating the quill epidermis.

Dust Bath.—Domestication of the hen can be carried so far and no farther, and this fact must ever be remembered. It might seem a little more sanitary or civilized for the hen to keep her body clean by using a white enameled bath tub provided with hot and cold water taps, or to have a chicken barber shop where a weekly shampoo may be had, but the hen prefers to wallow in the dust of the road or in a crude box filled with dust, that is not by any means germ-proof. It would be as easy to make water run up hill as to change the nature of the hen when it comes to her method of keeping the body clean. The dust shampoo rids the hen of the scurf of the skin; besides it is nature's protection against the pestiferous louse that has no object in life but to make living a burden to the hen. A dust bath at evening gives biddy a restful sleep; a dust bath during the day gives her new hope and happiness and permits her to lay her daily egg in peace and to chase and devour other larger insects that prey upon the crops of the field.

The hen louse must breathe to live, and it breathes through the pores of its skin. A knowledge of this simple fact was doubtless the clew for some ancient Edison to invent dust. There are some objections to dust, but there is always some bitter with the sweet, and to biddy dust tastes sweeter than plum jam to the average human youngster. The hen must have her daily dust bath. If she cannot get it in the fields or the yards it must be furnished in a box artificially, but she must have it to cleanse her body in the old natural way. It fills up the pores of the louse and prevents breathing, thus killing it.

The addition of sulphur, pyrethrum or lime to the dust makes it more effective on account of their irritating nature. If the poultry premises are kept in a sanitary condition, the fowls will keep themselves practically free from lice if they have access at all times to a good dust bath. By dusting the hen by hand with a good insect powder, the lice may be gotten rid of sooner, but this entails too much labor to be practicable on a large or commercial scale.

Sulphur and slaked lime may be used as a dust powder. Another good powder may be made by mixing crude carbolic acid 90 to 95% strength, with enough plaster of paris to make a dry powder.

Head lice on small chicks, which make their appearance a day or two after the chick is hatched, may be killed by rubbing the head and throat of the chick with lard. A few drops of kerosene to a teaspoonful of lard will make it more effective, but much kerosene may kill the chick. If the hen be carefully treated for lice while sitting, there will be less trouble from the head lice on the chicks.

Scaley Leg.—Another species of mite (*Sarcoptes mutans*), produces scaley leg in fowls. The mite burrows underneath the scales of the leg and white grayish crusts are formed which gradually enlarge and raise the scales. In severe cases lameness results and even the loss of toes. The disease is contagious. Disinfective measures should be applied in the poultry house. Individual treatment is rather tedious but a cure is easily affected. Where the case is bad or advanced it will usually be necessary to soak the scales thoroughly in warm water and remove them when it can be done without causing bleeding, then apply an ointment or vaseline. Kerosene is an effective remedy. Where the disease is not too far advanced it will be sufficient to dip the legs in a can of oil and hold them there for half a minute.

Dr. Theobald is authority for the statement that there are some 36 distinct species of worms that live as parasites in fowls. Some of these are of little importance. A few of the more injurious ones will be mentioned here.

The Gape Worm (*Syngamus trachealis*).—This parasite is very destructive to young chicks in different sections of the country. It attaches itself to the inner lining of the windpipe or trachea. Contaminated soil is responsible for the spread of this disease. This further emphasizes the point that has already been made, that young chicks should always be reared on clean, fresh ground. Gape worms become so numerous in the windpipe when they once get started that the bird finally dies for lack of air. Some of the worms are coughed up, as well as some of the ova and embryos and these are taken up by other fowls and the disease rapidly spreads. The ground carries the infection from one year to another. By cultivating the ground and disinfecting it with lime, the infection may be destroyed. Where the disease is known to exist it is the safe plan not to use the same ground for a year or two.

The chief symptoms are a gaping with open beak and stretching of the neck forward. The worm may be removed by twisting a horse hair in the windpipe and withdrawing it, or a feather stripped to near the tip, dipped in oil or turpentine may be used in the same way. The value of the young chicks, however, will not usually warrant individual treatment. Reliance must be placed upon keeping the chickens away from contaminated ground.

Intestinal Worms.—There are numerous worms that infest the digestive organs of the fowl. The round worms are found in the gullet or esophagus. Another species is found in the walls of the gizzard. The tape worms and various other species are found in the intestines. For individual treatment, oil of turpentine is recommended, one teaspoonful per fowl, preferably given in the morning followed with olive oil or castor oil a few hours later. Heavy feeding of onions or garlic will aid in controlling these parasites. Another remedy is to use powdered pomegranate root bark, one teaspoonful to 50 birds given in the feed.

There is not space in this book for an extended discussion of poultry diseases. Those readers wishing more detailed information of various diseases and their treatment, will find several books on this special subject. Among them may be mentioned: "Poultry Diseases and Their Treatment," by E. J. Wortley; "Diseases of Poultry," by Dr. D. E. Salmon; "Diseases of Poultry," by Pearl, Surface and Curtis; "Poultry Diseases," by Dr. B. F. Kaupp; and "Parasitic Diseases of Fowls," by Theobald.

INDEX

A

Albumin	281
Analysis of fowls and egg	229
Animal food	241
Ash	223

B

Balanced rations	219
Barley	239
Beef scrap	243
Beets	245
Beet pulp	246
Bran	239
Breed, Ancona	32
Andalusian	33
Black Spanish	33
Braekel	34
Brahma	55
Campine	33
Dorking	45
Faverolle	49
Hamburg	33
Houdan	34
Langshan	60
La Fleche	60
Le Mans	58
Leghorn	30
Minorca	31
Orpington	46
Plymouth Rock	39
Rhode Island Red	43
Sussex	49
Wyandotte	41
Breeding, principles of	61
problems in	62
purity in	68
Breeds, economic qualities of	27
edible meat on different	52
egg	30
fancy	28
general purpose	35
meat	50
origin of	30
"Standard" classification of	24
utility classification of	28
Broilers	274
Brood coop	198
Brooders, colony	324
home-made	324
lamp	323
stove or room	327
types of	323
ventilation of	322
Brooding, artificial	320
chicks, training	322
period	320
temperature	320
Buckwheat	240
Buttermilk	241

C

Cabbages	246
Capons and caponizing	372
Carbohydrates	226
Catarrh	382
Charcoal	247
Chickenpox	383
Cholera	383
Cockerels	273
Cold storage	359
conditions of	360
effect on prices of	359
limitations of	361
Colony system	144
Corn	237
Crop bound	385
Culling	272
Cut bones	241
Cross breeding	16, 73
advantages of	74
disadvantages of	79
experiments in	78, 101

D

Diarrhœa	383
Digestibility of foods	230
Digestion coefficients	232
Digestive organs	230
Diphtheria	384
Diseases	375
Disinfection	377
Domestication of fowls	1
purpose of	1
Dominance	84

E

Drainage 168
Drawn and undrawn poultry... 371
Dressing, loss in weight in.... 372
Dropsy 384
Dust bath 395

E

Edible meat on fowl.......... 372
Egg-laying organs 133
Egg, structure of............ 281
Egg production, limit of...... 112
    best in first year......127, 183
    progression in 109
    regression in 109
Eggs, candling 352
    canned frozen 361
    classification of 347
    color of 351
    composition of 217
    conditions that injure..... 348
    fertility and hatchability of 299
    for hatching 297
    fresh and stale........... 354
    gathering the 349
    grades of 344
    grading 351
    marking 355
    preservation of 362
    refrigeration of 357
    selling, for hatching..... 364
    size of 351
    testing 299
    washing 350

F

Fattening fowls 274
    batteries 277
    rations 278
Fats 226
Fecundity, inheritance of... 92, 134
    influence of sire and dam. 107
    Maine station's results... 93
    Oregon station's experiments 96
    sex limited 107
Feeding and exercise......... 249
    cooked food 254
    fundamentals of 210
    ground and unground grain 251
    growing stock 268
    hopper 254
    limitations of 212
    methods of 249
    purpose of 223
    rations 256
    small chickens 265

Fencing 205
    portable 203
Fish scrap 243
Food analyses 227
    animal 241
    carbohydrates and fat.... 226
    composition of 223
    computing the ratio of.... 229
    digestibility of 230
    digestion coefficients of... 232
    grain 237
    green or succulent........ 244
    mineral nutrients of...... 223
    palatability of 234
    protein 214
    relation of, to color of egg 212
        flavor of egg......... 212
        quality of eggs....... 212
        size of eggs......... 215
        yield of eggs......... 214
    requirements of chickens.. 221
Fowls, antiquity of........... 9
    evolution of 11
    origin of 2
Free range 144
Fruit trees for shade......... 206
Fruit growing and poultry-keeping 140
Fresh air, value of....... 174, 375
Fresh ground, preventive of diseases 376

G

Gallus bankiva 3
Gape worm 397
Green food 244
Grit 247

H

Hen-hatching 291
Heredity 62
Hygienic conditions, importance of 375
Historical 1
House, curtain-front 176
    floor of 184
    foundation 186
    open-front 186
    portable 187, 193
    space required 178
    stationary house 189
Housing, essentials of........ 160
    purpose of 167

INDEX

I

	Page
Inbreeding	86
Incubation, artificial	301
carbon dioxide and moisture in	311
chemical composition of chick influenced by methods of	316
cooling the eggs	316
influence of moisture in	312
loss of weight in eggs	313
methods of	286
moisture in	308, 309
natural vs. artificial	287
oil on egg shells	319
period of	297
temperature of	314
turning the eggs	316
wet bulb temperature as a moisture guide	309
Incubator, choice of	305
operating the	307
size of	307
types of	305
Incubator house	302
ventilation of	303
analysis of air in	304

J

	Page
Jungle fowl cock	3
hen	4

K

	Page
Killing	368

L

	Page
Lady Macduff	117, 119
Laying longevity	115
maturity	127
Lice	395
Limberneck	385
Linseed meal	240
Locations for houses and yards	167

M

	Page
Manure, preservation of	208
Marketing eggs	333
classifying eggs for	347
direct	337
express	338
how costs are added	336
indirect	335
poultry	366

	Page
Middlings	239
Milk	241
Milk albumin	243
Mites	393

N

	Page
Nests	196
Nutritive ratio	228

O

	Page
Oats	238
Oats and peas	246
Oregona	108
Oyster shell	247

P

	Page
Parasites	392
Parcel post shipments of eggs	340
poultry	371
Peas	239
Petaluma poultry farming	140
Peritonitis	384
Picking	369
Portable fencing and houses	203
Potatoes	246
Poultry farming, systems of	138
backyard	152
colony	144
dairying with	139
exclusive	152
fancy	156
fruit growing with	140
grain growing with	140
intensive	151
mixed husbandry	138
Petaluma	140
Rhode Island	140
specialized	138
industry	19
products	19
publications	19
Prepotency	83, 85
Preserving eggs in water glass	363

R

	Page
Rations	256
Reversion	65
Rheumatism	384
Rice	240
Roup	381
Rye	240

S

	Page
Scalding poultry	370
Scaley leg	397
Selection	14
Shade	206
Shaping	370
Shell, furnishes lime for developing chick	311, 316, 317
structure of	282
Soils	167
Sprouted oats	246
Sticking	368
Storms, objectionable	170
Sunflower seed	240
Sunshine	169

T

Transportation cars for poultry	367
Trapnests	198, 199
Tuberculin test	380
Tuberculosis	378
Type in layers	120

U

Use and disuse of parts	16

V

	Page
Variability	109
Variation	12, 63
factors influencing	13, 15
Ventilation in poultry house	174
Vetch and oats	245

W

Water glass	363
Weight correlated with laying	120
Wheat	237
White Diarrhœa	389
Worms, intestinal	398
Wry neck	385

Y

Yards	191, 201
crops for	192
cultivation of	191
double	203
hen capacity of	191
size of	202
Yolk	282

www.ingramcontent.com/pod-product-compliance
Lightning Source LLC
Chambersburg PA
CBHW062317220526
45469CB00008B/2543